T0258410

Re-Engineering the Manufacturing System

MANUFACTURING ENGINEERING AND MATERIALS PROCESSING
A Series of Reference Books and Textbooks

EDITOR

Ioan Marinescu
University of Toledo
Toledo, Ohio

FOUNDING EDITOR

Geoffrey Boothroyd
Boothroyd Dewhurst, Inc.
Wakefield, Rhode Island

1. Computers in Manufacturing, *U. Rembold, M. Seth, and J. S. Weinstein*
2. Cold Rolling of Steel, *William L. Roberts*
3. Strengthening of Ceramics: Treatments, Tests, and Design Applications, *Harry P. Kirchner*
4. Metal Forming: The Application of Limit Analysis, *Betzalel Avitzur*
5. Improving Productivity by Classification, Coding, and Data Base Standardization: The Key to Maximizing CAD/CAM and Group Technology, *William F. Hyde*
6. Automatic Assembly, *Geoffrey Boothroyd, Corrado Poli, and Laurence E. Murch*
7. Manufacturing Engineering Processes, *Leo Alting*
8. Modern Ceramic Engineering: Properties, Processing, and Use in Design, *David W. Richerson*
9. Interface Technology for Computer-Controlled Manufacturing Processes, *Ulrich Rembold, Karl Armbruster, and Wolfgang Ülzmann*
10. Hot Rolling of Steel, *William L. Roberts*
11. Adhesives in Manufacturing, *edited by Gerald L. Schneberger*
12. Understanding the Manufacturing Process: Key to Successful CAD/CAM Implementation, *Joseph Harrington, Jr.*
13. Industrial Materials Science and Engineering, *edited by Lawrence E. Murr*
14. Lubricants and Lubrication in Metalworking Operations, *Elliot S. Nachtman and Serope Kalpakjian*
15. Manufacturing Engineering: An Introduction to the Basic Functions, *John P. Tanner*
16. Computer-Integrated Manufacturing Technology and Systems, *Ulrich Rembold, Christian Blume, and Ruediger Dillman*
17. Connections in Electronic Assemblies, *Anthony J. Bilotta*
18. Automation for Press Feed Operations: Applications and Economics, *Edward Walker*
19. Nontraditional Manufacturing Processes, *Gary F. Benedict*
20. Programmable Controllers for Factory Automation, *David G. Johnson*
21. Printed Circuit Assembly Manufacturing, *Fred W. Kear*

Additional Volumes in Preparation

Re-Engineering the Manufacturing System

Applying the Theory of Constraints

Second Edition, Revised and Expanded

Robert E. Stein

Robert E. Stein Consulting
Cedar Park, Texas, U.S.A.

MARCEL DEKKER, INC.   NEW YORK • BASEL

Library of Congress Cataloging-in-Publication Data
A catalog record for this book is available from the Library of Congress.

ISBN: 0-8247-4265-6

This book is printed on acid-free paper.

Headquarters
Marcel Dekker, Inc., 270 Madison Avenue, New York, NY 10016, U.S.A.
tel: 212-696-9000; fax: 212-685-4540

Distribution and Customer Service
Marcel Dekker, Inc., Cimarron Road, Monticello, New York 12701, U.S.A.
tel: 800-228-1160; fax: 845-796-1772

Eastern Hemisphere Distribution
Marcel Dekker AG, Hutgasse 4, Postfach 812, CH-4001 Basel, Switzerland
tel: 41-61-260-6300; fax: 41-61-260-6333

World Wide Web
http://www.dekker.com

The publisher offers discounts on this book when ordered in bulk quantities. For more information, write to Special Sales/Professional Marketing at the headquarters address above.

To my wife, Debera, for her love and support;
to my daughters, Jennifer, Allyson, Elizabeth, and Hannah;
and to my granddaughters, Jessica and Samantha

Foreword to the Second Edition

In my opinion, more Texans should write books. Not only is the accent easier to understand, but they think big. Bob Stein thinks big. When he talks about re-engineering the manufacturing system, he's not talking about small changes or trivial refinements. That won't put you ahead in today's world. He's talking about fundamental changes in how manufacturing organizations do business and how their information systems support them. And he's talking about how your organization can make more money—a lot more money. Many of the ideas in this book will seem to be everyday common sense, but they are not common practice; and together they add up to a much-needed overhaul of manufacturing management systems.

The Theory of Constraints (TOC) asserts that any system has a very small number of limiting factors (constraints) that determine its performance. Every chain has a weakest link. This simple and obvious idea, in the hands of an expert like Bob, has tremendous practical ramifications. I am not exaggerating when I say that by implementing the concepts in this book, most manufacturing organizations will see more than 20% increases in throughput and 30% decreases in lead times, without significant additional expenses.

This means that "Theory of Constraints" is something of a misnomer. It is much too practical to be called a theory; that's a little like calling addition a "theory of aggregation." And while "constraints" do limit your results, that very fact means that focus on them is tremendously leveraged. Those of us who use the TOC concepts do so because they comprise a practical approach to figuring out where to focus a business's systems management to obtain maximal results.

Some of the ramifications of TOC to manufacturing systems have been addressed in earlier books (for the most comprehensive conceptual picture, see E.M. Goldratt, *The Haystack Syndrome*, North River Press, Croton-on-Hudson, NY, 1990). However, no one else has combined the practical knowledge of ERP systems with the practical knowledge of TOC to come up with this kind of authoritative, detailed manual. *Re-Engineering the Manufacturing System* provides an in-depth, soup-to-nuts description of everything they didn't teach you in MRP school: how manufacturing systems need to be structured in order to make money. And, speaking of money, I especially recommend careful reading of Chapter 14, which discusses a profit-based approach to implementing systems. This chapter by itself could save you years of heartache.

Bob is in a unique position to write this book. Besides being a Texan, he has 25 years of experience leading MRP and ERP implementations around the world. This has given him an intimate and detailed understanding of the traditional approach to manufacturing management, both the good and the bad. He has also spent many years studying, implementing, and developing Theory of Constraints concepts, in particular as they apply to manufacturing and manufacturing systems. Bob's background gives this book a knowledgeable, balanced perspective that will be especially appreciated by manufacturing professionals who, above all, value success.

Study *Re-Engineering the Manufacturing System*. Think big.

Rob Newbold
Chief Executive Officer
ProChain Solutions, Inc.
Lake Ridge, Virginia, U.S.A.

Foreword to the First Edition

With the publication of *Re-Engineering the Manufacturing System: Applying the Theory of Constraints*, Bob Stein has made an important contribution to the manufacturing systems portion of the emerging Theory of Constraints (TOC) literature. The book provides answers to two very basic and important questions in today's systems marketplace:

What should be re-engineered?
Answer: The infrastructure of manufacturing information systems.

What should it be re-engineered *to*?
Answer: To the public domain specification for integrated manufacturing decision support, shop control, and scheduling systems described in the book *The Haystack Syndrome*.

Bob is in an excellent position to write a book on this subject. With over 20 years of experience, he has led MRP implementations in large accounts throughout the world. His background in TOC includes attending the Avraham Y. Goldratt Institute's two-week TOC "Jonah" course and participating in the so-called second wave of the institute's Certified Information Systems Associate (CISA) program. Created by TOC systems pioneer Avraham Mordoch, the CISA program provided education and consulting support for the trailblazing manufacturing companies that participated as development partners with the institute. This period of conceptual development and field testing gave rise to the subject of this book, the public domain TOC manufacturing decision support, shop control, and scheduling specification described in *The Haystack Syndrome*. As a member of the American Production and Inventory Control Society (APICS) Constraints Management Special Interest Group (CM SIG), Bob has served as subcommittee chairman for "TOC and Total Quality" issues and courses. He is featured in the 1996 APICS National Workshops program as the developer and instructor for the course "TOC and Total Quality." That course arose out of Bob's timely and generous response to the CM SIG's request for donated APICS-owned TOC courseware and is based on his first book, *The Next Phase of Total Quality Management: TQM II and the Focus on Profitability* (Marcel Dekker, 1993). Currently, Bob is leading the CM SIG committee's efforts to formulate strategies concerning APICS TOC courseware and certifications in the area of TOC manufacturing systems.

The Theory of Constraints is a physics. It seeks to uncover the underlying nature of things. Its applications to the manufacturing industries are introduced in many publications, but primarily *The Goal* (1984, with a major and important 1992 revision), *The Haystack Syndrome: Sifting Information from the Data Ocean* (1990), and *It's Not Luck* (1994). *The Goal* and *It's Not Luck* are business texts written in the form of novels. Taken together, they have sold in the millions around the world in several languages. They are by far the most friendly and effective introductions to TOC and its major principles. The role of *The Haystack Syndrome* in the TOC literature is to define the generic solution for integrated manufacturing decision support, shop control, and finite capacity production scheduling. It asserts the need to replace traditional cost accounting for manufacturing business decisions, proceeds to develop the solution, and creates related specifications for shop control and scheduling along the way. The author of all three of these books is physicist and educator Dr. Eliyahu M. Goldratt, the founder of the Avraham Y. Goldratt Institute in New Haven, Connecticut, and the originator of the Theory of Constraints.

In the current volume, Bob Stein extends *The Haystack Syndrome* analysis in three important ways. First, he addresses several key shortcomings of existing systems, and several important aspects of the TOC solution, at greater levels of detail. For example, this book offers a more detailed look at the many *workcenter-level* "scheduling" tools available within traditional shop floor control systems and explains why there is simply no way they can perform as well as TOC's *company-level* drum-buffer-rope and dynamic buffering solutions. Not that the old system is bad; the traditional tools are in widespread use today and are helpful to many people in their daily work. However, Bob helps readers see that the TOC tools will allow those same manufacturing people to accomplish more on a larger scale with less effort from their daily "scheduling" activities. Second, he extends the *Haystack* analysis by tailoring the scope, terminology, and graphics to match what the manufacturing systems community normally sees in their textbooks. Finally, issues of data structure and implementation are addressed. These include changes to performance measurements, use of TOC thinking processes for dealing with policy constraints, and other matters which either were outside the intended scope of *The Haystack Syndrome* or were invented since its publication.

These contributions make *Re-Engineering the Manufacturing System* an especially helpful volume for manufacturing systems managers, practitioners, consultants, designers, and programmers seeking additional and timely information about the emerging TOC systems standards. Like his previous book on TOC and TQM, this text will assist many disparate constituencies in taking important initial steps toward understanding, consensus, and cooperation.

Thomas B. McMullen, Jr.
Founding Chairman
APICS Constraints Management Special Interest Group (CM SIG)
Weston, Massachusetts

Preface to the Second Edition

New software that provides help to manage customers and vendors and to optimize entire supply chains suffers from the same malady as the original enterprise resource planning (ERP) systems. They were not designed after a complete analysis of the environment in which they were to work. It is synchronization that is assumed by all factory scheduling or supply chain systems. It is synchronization that is required to maximize profit. However, synchronization demands that the physical laws that govern the environment be obeyed. If they are not, synchronization—regardless of how much the user or designer might want it—will be impossible. Unfortunately, it would come as a complete surprise to most designers that optimization and synchronization are not necessarily the same things.

In addition to the systems that have been developed, implementation strategies, while they have undergone some degree of change, have yet to be successful in living up to their potential. In fact, if you ask most CEOs, you will find that implementing ERP systems can be like a bottomless pit into which money is poured. Little, if any, benefit ever reaches the bottom line. Implementation strategies, like information systems, must change so that they focus on profitability, and implementers must be held accountable for their results.

The second edition of this book is focused, in part, on the concepts of synchronization versus optimization and points out the obvious differences between them. Additionally, it presents a paradigm shift in implementation strategies focusing not on application requirements or process improvements, as most implementations are, but on profit requirements. Taken in a phased approach, implementations of ERP and supply chain systems should increase profits significantly and reach the break-even point before the project is complete.

Again, I would like to offer my thanks to some special people who devoted their time to making this book much better than it would have been otherwise. Thank you, Dr. Eli Goldratt and Rob Newbold for your review recommendations. Thanks also to the staff of the ProChain Corporation for their help in clarifying critical chain project management systems.

Robert E. Stein

Preface to the First Edition

Originally designed by software engineers to fit the computer environment, traditional manufacturing system concepts are now over 30 years old. In that time tremendous changes have occurred in hardware and software, but little has changed in the basic design concepts. The processes of master production scheduling and gross-to-net requirements generation are the same now as when first conceived. Little progress has been made to really improve the majority of systems being marketed today. Recent trends that increase the rate at which planning occurs to simulate scheduling or materials planning scenarios have merely created the opportunity for users to do the wrong things faster.

The traditional closed loop systems were not designed by examining the cause-and-effect relationships within the environments in which they were to function or by developing cold, hard facts about how resources are related to one another. Had the original systems been based on a thorough understanding of the physical requirements rather than by looking first at what the computer could do, a totally different system would exist today. It would be a system that was fully grounded in reality. It is only when the problem has been sufficiently defined that a system can be created which will truly solve the problem.

In 1990 Dr. Eliyahu M. Goldratt presented an outline of such a system in his book *The Haystack Syndrome*. Also included were a measurement system and the thought processes behind why the system must work as it does. In recent years the "Haystack Compatible" system has been growing in acceptance and has even been promoted by the recently established Constraints Management Special Interest Group (CM SIG) of the American Production and Inventory Control Society. This is an important change. One hopes it means that the old MRP-based systems are going to be replaced by ones that discriminate between data and information.

If the decision process is to change to the extent advocated, then it is time to take a serious look at making some changes in the information system—not superficial changes like adding additional reporting functionality, but major ones. This book is a detailed instruction guide on how to build, implement, and use a manufacturing system based on the Theory of Constraints (TOC). I hope to give enough information so that manufacturing software companies can begin the process of modifying the traditional system to better conform to reality and to help those faced with the job of implementing a system to better understand what they are attempting to do.

I would like to offer my thanks to some special people. Without their education, patience, and dedication to a worthy cause, this book would not have been possible. Thank you, Dr. Eli Goldratt, Avraham Mordoch, Eli Schragenheim, and the staff and former associates of the Goldratt Institute.

Robert E. Stein

Contents

1
An Introduction

This chapter offers a brief history of the manufacturing information system, introduces the possibility that an improved methodology is available, and lays the groundwork for what is to follow.

I. CHAPTER OBJECTIVES

- To provide an overview for this book
- To lay a historical foundation

II. THE ADVENT OF THE TRADITIONAL SYSTEM

In 1974 Joseph Orlicky documented in his book *Material Requirements Planning* what had become one of the biggest changes to American manufacturing since the Industrial Revolution. Orlicky presented in intricate detail the design, interface, and calculation issues for what was called the manufacturing information system.

The advent of the computer and its ability to store and manipulate large volumes of data made it possible to create demand for upper level finished goods items, drive demand for lower level assemblies and raw material parts, and schedule the shop floor. For an industry that had been planning every part based on individual usage rates this was a major step forward.

While he was not alone in this change, his book would become one of the most read in the American Production and Inventory Control Society (APICS) library for years to come. Oliver Wight and other pioneers in the field had also begun this documentation process. Each had been involved for some time in the creation of these systems. During the late 1960s and early 1970s, as these systems became reality, companies would see an opportunity to mainstream the computer system into American business. By the end of the 1970s and early 1980s the processes of master production scheduling, material requirements planning, capacity requirements planning, and shop floor control became commonplace. The closed loop system (Fig. 1) became the most recognized graphic of the manufacturing systems industry.

The accepted technology had become so ingrained in industry, software companies, and consulting organizations as well as at the university level that it soon became very difficult to make any significant changes. Those changes that did occur were simply built on top of the assumptions upon which the original systems were designed.

Hampering any efforts to create improvements in the technology were the millions of dollars invested to create the software that would run this nation's economy. Once written, manufacturing systems became not just software, but an investment that had to be protected by those who wrote or bought them. And, once learned by the user, this knowledge became job security. The slowness of the acceptance of techniques like just-in-time manufacturing or the Theory of Constraints (TOC) is evidence of the difficulties that had to be overcome in implementing new methodologies. Possibly the greatest problem to overcome can be summed

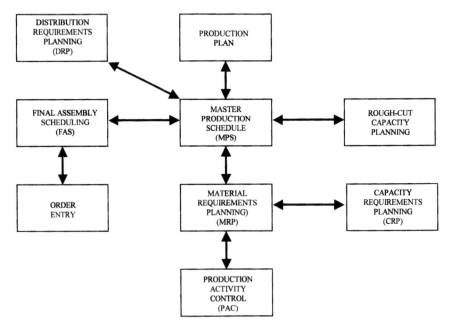

Figure 1 The closed loop system.

up in a statement uttered by the head of manufacturing consulting for a major U.S. corporation, "It doesn't matter whether you're right or wrong this is what General Motors wants." There is a tremendous amount of money to be made in the consulting industry in this country. For the benefit of the uninitiated it must be understood that being right isn't necessarily the goal.

It was never a secret that there might be problems with the technology. In Orlicky's book, now 20 years old, pages 46–48 provide some insight into the conflicts that existed at the time and are made very clear. Material requirements planning (MRP) was known to be "capacity insensitive." However, the benefits so far outweighed the problems inherent in the order, launch, and expedite system commonly known as EOQ-ROP, or economic order quantity and reorder point, that the problems were ignored. At least for a while. The advent of finite capacity planning seemingly made it possible to now synchronize the way in which raw material was ordered with the way it was to be consumed by the resources and the market. Unfortunately the problems involved in attempting to accomplish this task were based upon the assumptions of the original system, making it virtually impossible to do so effectively.

Two issues that greatly hampered the creation of an effective system were cost accounting and a complete lack of knowledge of the physical laws that govern the manufacturing environment. The objective of the original system was to use the computer's calculating abilities to time phase material requirements. However, to do this correctly meant that the system must resemble reality. And this is where things were done that made today's manufacturing systems much less effective than they could be.

III. EXPLORING A DIFFERENT APPROACH

In the mid- to late 1970s Dr. Eli Goldratt, who had a background in physics, approached the information system from a totally different perspective. His observations, unfettered by the traditional thought process, resulted in his solving some very interesting problems and creating a technology that has proved to be far superior to the traditional manufacturing system. In his books *The Goal* (1992), *The Race* (1986), and *The Haystack Syndrome* (1990) and in classes at the Goldratt Institute, Dr. Goldratt presented a totally new way of looking at information as being the answer to the question asked. While this seems a trivial issue and is just common sense, if you get the question wrong or do not know how to answer it, you cannot begin to build the information system. This is where the fun begins. The questions as well as the answers are totally different than what had been taught for the last 50 years. With this in mind, how could anyone be expected to build the information system?

A. Providing the Key

It is very clear that a manufacturing system is not just a tool for planning material and scheduling the factory. It is an integral part of the five steps of improvement for the "throughput environment" as defined in *The Haystack Syndrome* and in *The Theory of Constraints: Applications in Quality and Manufacturing* (1997). The key issue that defines the effectiveness of the information system is how well it supports the TOC principles of

1. Identifying the constraint
2. Exploiting the constraint
3. Subordinating the remaining resources
4. Elevating the constraint
5. Repeating the process

It is this sequence of steps that literally defines what an information system must do. Following them to their logical conclusions means that a system must aid in the identification, exploitation, and elevation of the constraint and with the subordination of the remaining resources. For this to occur means that the logic of the system must be very different than it is in the traditional system.

In *The Haystack Syndrome*, Dr. Goldratt also defines the structure of the information system. He establishes three separate segments along with justification for each. These segments include

- What if
- Control
- Schedule

His objective was to create a system which can help in answering questions, or to play the "what if" game. To support the "what if" there must be a "schedule" so that the physical laws governing the process can be addressed, and there must be a "control" so that problems that may arise can be addressed before they impact the schedule. This whole concept will be addressed in Chapter 3.

IV. AN OVERVIEW OF THIS BOOK

The remaining chapters of this book are dedicated to presenting the problems associated with the traditional system's ability to support the five steps based on the TOC principles and to present, in detail, the solutions provided by the Theory of Constraints. The objective is to provide enough background as well as detailed information to allow industry to begin the process of redefining the information system.

It represents a tremendous change in the distinction of what an information system is and what it should do and lays the groundwork for the design of the next generation of systems, including

- Design
- Focusing mechanisms
- Processing logic
- Structure

The objective here is not to eliminate the MRP II system but to change it so that it will be more in line with the goal of the company. MRP II has created a great data system upon which to draw. How to use these data to implement the five steps is another issue. There must be a change.

A. Chapter Topics

- **Chapter 2: Visiting the Traditional System** is a general over-view of the problems associated with the traditional manufacturing system. The traditional system presents some very serious problems that must be overcome before it can really be called an information system.
- **Chapter 3: Synchronization Versus Optimization.** The recent introduction of advance planning and scheduling systems and supply chain management systems has only served to further confuse the market as to what is necessary. Chapter 3 presents a comparison between systems that serve to optimize and those that attempt to synchronize and shows why synchronization is a preferred solution.
- **Chapter 4: Establishing the Characteristics of an Information System** addresses from a conceptual perspective what an information system should do and how it should be organized. This is the first step in building a system that is based on reality.
- **Chapter 5: Defining the Structure** provides a global view of how the information system is structured. It puts every segment of the system into place and defines the interfaces between the various processes, including input and output.
- **Chapter 6: Identifying the Constraint** deals with the intricate details of how the system identifies the constraint, including processing logic.

- **Chapter 7: Exploitation.** It is not enough to identify the constraint. It governs the amount of money a company can make. It must be exploited for every dollar, yen, or pound that can be gained. Chapter 7 presents the processing logic for the exploitation phase.
- **Chapter 8: Subordination.** Once the constraint has been exploited, all the remaining resources must give the constraint what it needs and nothing more. Chapter 8 details the subordination logic.
- **Chapter 9: Buffer Management.** The system must be protected from those things that can and will go wrong. Chapter 9 presents the logic of the buffering process and buffer management.
- **Chapter 10: Supporting the Decision Process.** Once a complete definition of an information system has been made, how then should it be used to maximize the creation of throughput? By aiding in the process of answering specific questions, such as what lot size should be used, should a certain product be obtained from internal or external resources, and what price should I accept for a given product, the system can provide information that will make a difference.
- **Chapter 11: Making Strategic Decisions.** How and where a company should grow is important. One of the most significant uses of the information system is in setting the strategic direction. Chapter 11 presents how the system should be used to segment markets, plan corporate growth, plan long-term efforts, and recession-proof the company.
- **Chapter 12: Modifying the Current System** addresses the issues associated with how to modify the current system to conform to reality. It answers the questions of what should be eliminated and what should be kept. Throwing out the traditional system is not the answer. The current information system can provide a tremendous platform on which to build. What should be kept and what should be eliminated are major questions that need to be addressed.
- **Chapter 13: Implementing a TOC-Based Information System.** When the design of the system is shifted to the extent that is being addressed here the old methods often can cause problems. Implementing a different kind of system can mean using a different approach. Chapter 13 addresses those changes that must be made as the new information system is implemented in order to be successful.
- **Chapter 14: Implementing an ERP System to Improve Profit.** To be effective, an implementation of an information system must result in achieving the goal of the company. Chapter 14 extends the concepts established in chapter 13 to include the ERP system

implementation. It provides structure and direction to change what is termed an application-based implementation to a solutions-based implementation.

- **Chapter 15: Critical Chain Project Management.** How projects are scheduled and where to place management emphasis are critically important to insuring that projects are completed on time and on budget. It is also critically important to insure that projects are accomplished in the shortest possible time. Chapter 15 addresses the issues involved with establishing and maintaining project synchronization. Additionally, it presents an improvement process designed to shorten project life cycles.
- **Chapter 16: Scheduling the Multiproject Environment.** Multiprojects increase contention between resources and present additional challenges to the scheduling process. Critical chain multiproject capabilities minimize resource contention between projects while maximizing management visibility.

V. SUMMARY

Re-engineering the traditional system presents some interesting opportunities for improving the profitability of many companies. While it is not the only issue which must be addressed, it can make a major difference not only in scheduling the factory, but also in helping to eliminate many managerial, behavioral, logistical, or policy constraints that limit the company's ability to "make more money now as well as in the future." The technology is available. It is time to realize that a change can and should be made.

VI. STUDY QUESTIONS

1. What was the impact that the traditional information system had on manufacturing companies in the United States, and what benefit(s) does it provide?
2. According to the theory of constraints what process must be embedded in the logic of a successful information system.
3. What is the definition of information?
4. Describe the traditional closed loop system and provide a diagram indicating the order in which individual modules should appear.
5. What is the basic structure of the information system?

2
Visiting the Traditional System

The basic problem associated with the traditional system, whether of infinite or finite capacity, is that the processing logic used makes it virtually impossible to successfully implement the five steps of improvement. The user is unable to identify or exploit the constraint and to subordinate the remaining resources. It was just not designed for this purpose, and yet this is what must be done. (See Chapters 6, 7, and 8).

I. CHAPTER OBJECTIVES

- To address the problems associated with the traditional MRP II system, including rough-cut capacity planning, material requirements planning, capacity requirements planning, and shop floor control
- To begin the process of understanding the problems that the new generation of systems must solve

II. THE ASSUMPTIONS OF THE TRADITIONAL SYSTEM

The processing logic of the traditional system makes some very large assumptions which reality will not ignore. Those assumptions include but are not limited to the following topics:

- Relationships between resources
- Static versus dynamic data
- The aggregation of demand
- Statistical fluctuations

A. Relationships Between Resources

Because resources are not just isolated islands of production but interact regularly, they will have specific influences over each other based on where they are in the process and their physical characteristics. These influences must be reproduced in the system. The information system must be a simulation of reality. Without this quality, the information system is merely a data system.

B. Static Versus Dynamic Data

Hampering any ability to simulate reality within the company is the way in which certain data are used. Data that should be dynamic in nature are instead stored as static and recalled from the database without consideration of the changes which may have occurred or the physical laws which govern how it should be used.

C. The Aggregation of Demand

During the process of assessing available capacity and comparing it to demand, the tendency is to aggregate both into time buckets. This process overlooks one very important issue. It assumes that all the time in a given time bucket is available for all orders in the bucket. Unfortunately this is not a valid assumption and is a major issue with regard to how the traditional system manages capacity.

D. Statistical Fluctuations

In manufacturing, as in nature, all things will vary. This is a basic law. No two items are produced exactly alike. They will always be different. Similarly, manufacturing resources will always fluctuate in their ability to produce. Sometimes they will produce more than average. Other times they will produce less. The methodology for handling this in the typical manufacturing system is to build in buffers between each resource. It is not only how these buffers are built which raises some interesting questions, but how the instructions for executing the production plan cause problems by providing data which will sabotage the entire scheduling process.

   Note: Since each of these issues applies throughout the system, the approach used will be to describe the traditional system and then apply the assumption involved with the preceding topics. The most logical place to begin is with the master production schedule.

III. MASTER PRODUCTION SCHEDULING

The objective of the master production schedule (MPS) is to provide input to the system for those requirements currently needed by the customer (sales orders) or those that have been predicted to be needed in the future (forecast). It is significant in that it is what drives the remaining processes within MRP II (see Fig. 1).

From the master schedule, requirements are driven into material requirements planning (MRP) so that the quantity of specific components, lot sizes, and their time phasing can be identified. From MRP action messages are created to launch purchase orders for parts which must be obtained from outside the company and to launch work orders for those items to be built inside the company. Using the input of the MPS, MRP becomes the guiding factor in creating the schedule for the entire facility.

Given this information, it is obvious that to be successful in creating a valid master production schedule the ability of the system to produce according to demand requires that the capability of the system be known.

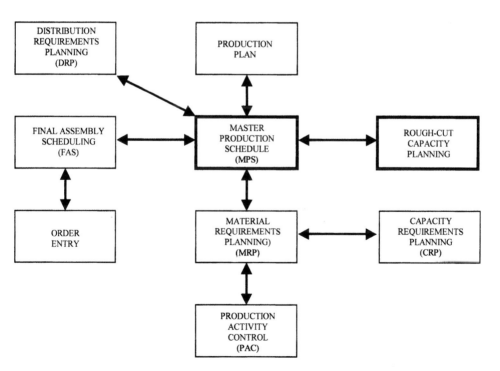

Figure 1 The closed loop system—master production schedule.

Part: AFG - Planned Routing

Work Center	Time/ Part
R-1	3.0
R-2	5.0
R-3	2.0
R-4	4.0

Figure 2 The product profile.

The traditional method for the process of understanding the capability of the system prior to attempting to create the MPS is called rough-cut capacity planning (RCCP).

A. Rough-Cut Capacity Planning

To provide the ability to perform rough-cut capacity planning many systems will store a profile of how certain products are produced. Included in the profile is the related load to be placed on specific resources during the manufacturing process.

Figure 2 represents a specific routing or load requirement to produce one part AFG. Stored for each resource, R-1 through R-4, is the time required for each step (Fig. 2). Notice that it takes 3 hours on resource R-1 to produce one AFG. These data when combined with demand from the MPS and similar data for other parts on the master schedule will create a total demand and percent load figure for all resources in the process. Figure 3 presents demand for ten AFGs in period 4 originating from the master schedule.

Part: A	1	2	3	4	5	6
Requirement				10		

Figure 3 Establishing demand with ruff-cut capacity planning—ten parts shown in period 4.

Work Center	PRODUCT			Capacity	Demand	Load %
	AFG	DEF	XYZ			
R-1	30	50	60	120	140	117
R-2	50	20	30	100	100	100
R-3	20	30	20	130	70	53
R-4	40	40	10	120	90	75

Figure 4 Rough-cut capacity planning example data.

This demand is multiplied by the individual resource time to produce the part (see Fig. 4). The demand for ten AFGs in period 4 of the MPS created a load of 30 hours on resource R-1. When combined with the demand for DEF and XYZ the total demand for R-1 is 140 hours, or a 117% load.

1. Time Phasing the Rough-Cut Capacity Plan

Some systems will take one extra step and attempt to time phase by storing relative lead time with the load data for each resource.

Figure 5 presents five operations each being processed on different resources. It also shows setup and run times as well as the relative period to load the demand for each resource. When the additional lead time data are added it facilitates placing each operation in relative time with the next. Notice that operations 10 and 20 fall into period 1 while operations 30 and 40 fall into period 2. After being extended by the quantity and due date for the individual product on the MPS, the load requirements for setup and run time can be placed in a relative position in time (see Fig. 6).

The ten parts required from the MPS for period 4 have been converted to labor requirements and placed in time for each of the resources in the load profile. Note that the labor for the saw and lathe appear in period 2, the first day of processing. Mill and heat treat were placed in the second day of

LOAD PROFILE

OPER.	WORK CENTER	SETUP TIME	RUN TIME	PERIOD
10	SAW	.15	.50	1
20	LATHE	.50	1.50	1
30	MILL	.30	1.30	2
40	HEAT TREAT	.50	1.50	2
50	INSPECT	.15	.50	3

Figure 5 Time phasing the load profile.

Master Schedule

Part: A	1	2	3	4	5	6
Requirement				10		

Load Requirements

W/C	1	2	3	4	5	6
Saw		5.15				
Lathe		15.50				
Mill			13.30			
Heat Treat			15.50			
Inspect				5.15		

Figure 6 Time phasing rough-cut capacity plan.

processing. The load for the inspection operation appears in period 4, the last day of processing.

2. Addressing the Problems

Even when lead time is added the data created in no way resemble reality. It is a gross distortion to assume that all time within a given time block is equally available to each order.

Figure 7 shows a capacity plan with a level load for five periods. From this illustration it is apparent that this resource has no real capacity problems. However, this is an oversimplification of what really happens as orders are placed in time.

The objective of capacity planning is to understand the impact of a specific load on the system's ability to deliver to the customer. Aggregation of demand into time blocks as in Fig. 7 hides the fact that each order is placed in a specific location in time with individual start and stop dates, and this placement has certain implications.

Figure 8 shows five orders, A through E, on a single resource and is a more realistic view of how orders are scheduled across a resource and why the aggregation of demand causes problems.

Each order is placed in time so that it has adequate protection and will arrive at shipping by a specific scheduled due date. By aggregating demand the assumption is made that the time to the right of order D is as equally available as the time to the left. However, the time to the right exists to protect the delivery to shipping on time. If any of the load for D is moved to the right, the sales order due date will be immediately invalidated. The only additional time available to process order D is to the left. But, under rough-cut capacity

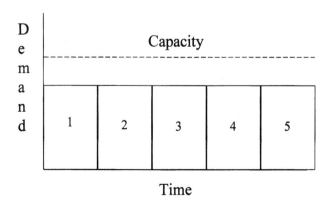

Figure 7 The aggregation of demand.

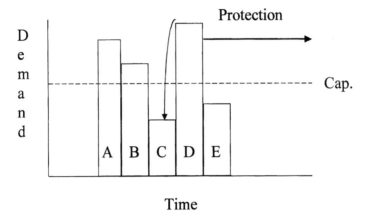

Figure 8   Offsetting for protection.

planning all time is considered equally available within the same time block. This is a serious problem.

Aggregating demand to a given time bucket is like putting one hand in boiling water and the other in a block of ice and feeling OK because on average you're at 98.6 degrees Fahrenheit. Unfortunately, a rough-cut is not good enough to be able to set the master production schedule. This means that the initial estimation of material requirements based on the input of the master production schedule would have no basis in reality. It is simply a wish list.

Since rough-cut capacity planning has failed to provide adequate capacity information, the hope at this point would be to set a preliminary master schedule, input that data to material requirements planning, and then check to see if capacity is available through capacity requirements planning (CRP). Unfortunately, the logic of MRP and CRP also presents major problems in accomplishing this task.

B.   Material Requirements Planning

Material requirements planning has been called the engine that makes the manufacturing system work. It takes output from the master production schedule and

- Determines the parts requirements for lower level assemblies and raw material
- Manages the timing of planned and scheduled work orders and purchase orders

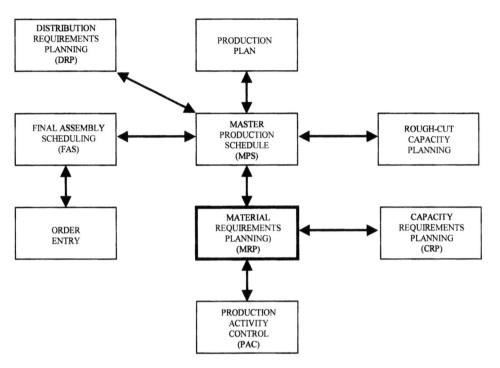

Figure 9 The closed loop system—material requirements planning.

- Notifies the materials planner when to launch new orders or reschedule old ones
- Drives the capacity requirements planning and shop floor control systems

Because of its position in the overall manufacturing system it is an extremely important process. However, it is also one which has some very serious problems (see Fig. 9).

1. Dynamic Versus Static Data in MRP

For an information system to resemble reality it must be dynamic in nature. While MRP II systems are dynamic and are able to change as the situation changes with respect to issues such as the planning of material requirements, in general they do not have the kind of dynamics required to recreate and react to those physical laws which govern the environment in which they operate. This issue alone makes it impossible to use MRP in executing the five steps of improvement. As an example, certain items of data such as lead times are

	1	2	3	4	5	6
GROSS REQMTS.						10
SCHED. REC.						
PLAN O/H						
NET REQMNTS.						10
PLAN REC.						10
PLAN REL.				10		
LEAD TIME = 2 LOT SIZE = L4L						

Figure 10 Gross-to-net requirements generation.

stored in the item master as static, yet lead times should grow and contract as the dynamics of the situation dictate. In the gross-to-net requirements matrix shown in Fig. 10 there are gross requirements of 10 As due in period 6.

There are no scheduled receipts or on-hand quantities to offset the need. So net requirements of 10 As are generated for the same period. Since the lot size to be used is lot for lot, as seen in the item master data at the bottom of the figure, 10 As are planned to fill the net requirement. The lead time from the item master is used to determine the release date of the order. The quantity of 10 As to be received in period 6 generated a planned order for 10 As to be released in period 4. In this case, MRP assumed that both the lead time and the lot size were static and would not change.

An issue can be made that most systems use the concept of fixed and variable lead time to solve this problem. As illustrated in Fig. 11, the planned receipt was doubled resulting in a corresponding increase in lead time. The original order planned for period 4 is now due in period 2.

This issue serves to illustrate the lack of understanding on the part of MRP developers. The key question that must be answered is what generates the need for a specific lead time? To understand this also means understanding what causes lead times to fluctuate. It is possible for two significantly different lot sizes under similar circumstances to have no difference in lead time requirements. Why? It is also possible for the same lot size under different circumstances to have a large difference in lead time requirements. Why?

	1	2	3	4	5	6
GROSS REQMTS.						20
SCHED. REC.						
PLAN O/H						
NET REQMNTS.						20
PLAN REC.						20
PLAN REL.		20				

LEAD TIME = 2
LOT SIZE = L4L
LEAD TIME = Variable

Figure 11 Variable lead time.

To answer these questions requires a better understanding of capacity. There are three types:

1. Productive—the capacity required to produce a product.
2. Protective—the capacity required to offset for those things that can go wrong.
3. Excess—the capacity which is over the amount of productive and protective capacity required.

It is the fluctuation in the protective capacity available that determines lead time. Figure 12 illustrates this.

As the load on a given resource increases, the amount of capacity available to use in the protection of the constraint actually decreases. At a 70% load, resources A, B, and C have enough capacity to ensure that

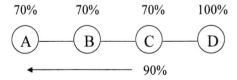

Figure 12 Fluctuation in protective capacity.

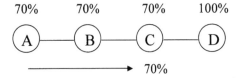

Figure 13   Fluctuation in protective capacity.

resource D (the constraint) is protected from any delays that might occur. However, if the load on resource C were to be increased to 90%, the amount of capacity available to protect the constraint would decrease. In other words, there would not be enough capacity left at resource C to ensure that if something were to delay production for any reason orders would still reach D on time. To ensure that inventory reaches the constraint without excessive delay raw material must be released earlier, thus creating an increase in lead time.

As the load decreases and the available protective capacity increases, the fluctuations in the productive capability have less of an impact so that material can be released later, thus causing lead times to shrink (see Fig. 13).

When resource C's load is returned to 70% the amount of protection for D is increased and raw material can be released later. This means that in order to accomplish the process of offsetting for lead time there must be some estimation of the amount of protection required while running the material requirement process. This is a different situation than what is being answered by the rough-cut or capacity requirements planning process.

As illustrated in Fig. 10, gross-to-net requirements generation considers lead time as static. In Fig. 11 it assumes that an increase in lot size would automatically increase the load and therefore increases lead time without ever looking at the impact on protective capacity. Even though the lead time seemed dynamic in the second example, it was really just a multiple of static data stored in the item master and had no basis in reality.

2.   The Cascading Effect

The problems associated with static data are exacerbated when it is considered that each part within each level of the bill of material is treated in the same fashion (see Fig. 14).

In gross-to-net requirements generation, the planned release of the parent becomes the gross requirement of the child, and the netting process continues at the child level. The item master is once again consulted for lead time and lot sizing information. Notice that the end item, product A, has a 2-day lead time. Part B's requirements are created to fill parts requirements

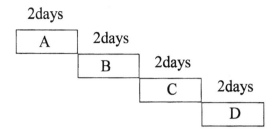

Figure 14 The cascading effect.

for an order of As. But part B must be available in its entirety to fill A's material requirements at the start of the order. This means that regardless of the reality of what might be occurring on the shop floor or what might be necessary to compress lead time, MRP ignores it.

In Fig. 14 MRP has already assumed that the lead time to create product A is 8 days by simply adding all of the static lead times in the process. Anyone who has ever implemented an MRP system has run into problems resulting from this issue. Lead times are exaggerated and inventory goes up.

Additionally, material requirements planning assumes that the process batch size and transfer batch size will remain the same throughout the process. It does not plan for overlapping operations. If the best method for reducing lead time is to transfer in batches of one, then MRP must plan its work orders in batches of one—not a good idea.

3. The Use of Lot Sizing Algorithms

Most MRP systems have a number of lot sizing methods used to sub-optimize between the cost of setup and the cost of carrying inventory. Various methods include but are not limited to

- Economic order quantity
- Part period balancing
- Period order quantity
- Least unit cost
- Least total cost

For these methods the thought process is that the higher the number of setups, the greater the setup cost. Conversely, the larger the lot size, the larger the carrying cost. Supposedly there is a balancing point where the minimum total cost can be determined by choosing a lot size which sub-optimizes the setup and inventory carrying cost. If the lot size is any larger or smaller than the economic order quantity, then total cost will be driven upward.

    a. Economic Order Quantity. The economic order quantity (EOQ) formula is stated as follows:

$$\sqrt{\frac{2US}{IC}}$$

where
U = Annual usage
S = Setup cost
I = Inventory carrying cost
C = Unit cost

The EOQ formula is represented graphically in Fig. 15.

    Line AB represents the declining setup cost as the lot size becomes larger and fewer setups are actually made. Line CD represents the rising carrying cost associated with having to carry more inventory for a longer period of time. Where the two lines meet is the point at which the total cost is the lowest. The lot size of 50 would require that five setups be made to fill an order for 250 parts. A lot size of 250 would only have one setup.

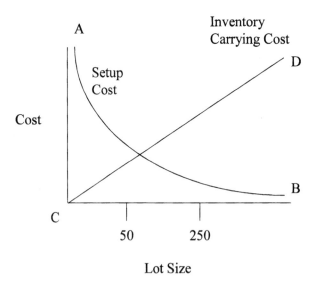

Figure 15 The economic order quantity.

The concept of the economic order quantity makes some very questionable assumptions. If any one of these assumptions is broken, any economic order lot sizing process would be invalidated, including all those previously listed. The assumptions made are

1. Each setup activity will result in a cost since activity absorbs labor dollars.
2. Setup costs will remain constant.
3. A given lot size will remain constant.
4. Carrying cost will remain constant.
5. Products have costs and those costs will remain constant.

Assumption 1 begs the obvious question what would be the impact if the five additional setups required by the 50-part lot size from Fig. 15 were made with excess capacity? If this were the case, there would be no difference in the cost of the setup for each lot size. This an important issue. Additional setup does not necessarily create additional expense. Since the objective of the EOQ formula is sub-optimization between the cost of setup and the cost associated with carrying inventory we need to go no further. The formula is unreliable. It cannot predict the actual impact of various lot sizes from an economic perspective. It does not consider the impact of capacity in the calculation of the lot size or on the profitability of the company.

Assuming that setup costs will remain constant for a particular lot size (assumption 2) ignores the fact that in some cases the additional setups will be made on resources which have excess capacity. Other times the exact same lot size and part will be made on resources which use protective or productive capacity. In the later case, the cost incurred is a result of either an increase in operating expense due to overtime, an increase in inventory due to parts being released earlier, or a decrease in the creation of throughput.

Assuming that a specific lot size will remain constant (assumption 3) is also questionable given that any lot may be split multiple times during the manufacturing process. The EOQ formula assumes that the process and transfer batch will always remain the same.

Assuming that carrying cost will remain constant (assumption 4) or that products have cost (assumption 5) presupposes that a particular rate of cost can be predicted or that the cost of a product even exists in order to create the carrying cost figure.

The whole concept that products cost a certain amount stems from the idea that this information can be used to make decisions. If decisions cannot be made based on the cost of a product, then cost as an entity becomes irrelevant. It is merely an illusion. When the amount of money spent to gather cost data or the economic impact of the poor decisions that result from it are considered, it becomes a very expensive illusion to maintain.

Cost accounting was created 100 years ago to solve problems in an environment that no longer exists. Nowhere in cost accounting is the concept of capacity discussed, and yet this is the overall governing factor that must guide the decision process. See Chapter 5, "Correcting the Decision Process," of *The Theory of Constraints: Applications in Quality and Manufacturing* for further background discussion.

b. Additional Considerations. Other lot sizing techniques not attempting to sub-optimize but which are used in MRP include

- Lot for lot
- Fixed order quantity

While these techniques do not suffer from the illusions of reducing cost through lot sizing, they still represent static data and make the assumption that the process and transfer batches must be the same.

These are issues which are not necessarily caused by the lot sizing algorithm chosen, but rather by the way MRP works. During gross-to-net requirements generation no attempt is made to determine what lot size will maximize throughput at the constraint, yet this is a very important step within the five-step process.

c. Conclusion to the Lot Sizing Problem. Since the governing methodology behind the information system is to help in the overall implementation of the five steps of improvement, any discussion of lot sizing must begin there. The key issue must be the impact of a given lot size on the exploitation of the constraint and the ability for the remaining resources to subordinate to the way in which it has been decided to exploit the constraint. It is only in the attempt to actually accomplish this task that the right lot size can be found (see Chapter 7).

Additionally, the true test to determine success or failure of a specific lot sizing strategy will be the impact of the strategy on the three measure of throughput, inventory, and operating expense. Will there be a change to these measure as a result of a specific lot size? The key questions to ask in order to solve the lot sizing problem are

- Will the particular lot size make inventory go up or down?
- Will there be a need for more or less operating expense?
- Will throughput be enhanced or will it decline?

If an answer cannot be determined for these three questions, then a satisfactory lot size cannot be determined. The question becomes what are the factors in the manufacturing environment which impact a given lot size?

C. Capacity Requirements Planning

Capacity requirements planning (CRP) takes the output of MRP and, by using the work center and route files, time phases the order in which specific activity is to take place on a given set of resources. The objective is to ensure that adequate capacity is available to meet the detailed schedule of material created by MRP (see Fig. 16).

After obtaining the output of MRP (Fig. 17), capacity requirements planning reads the route file (Fig. 18) to obtain the demand requirements for specific parts at their respective operations and work centers. This time is then multiplied by the amount of the order and then placed into time buckets (Fig. 19).

In Fig. 17, 10 part As are planned for release in periods 2, 3, and 5. According to the routing, each part will consume 1.0 hour of assembly time and each order will have 1.0 hour of setup. With a lead time of two periods, the 11 hours total time for each order is placed in periods 3, 4, and 6.

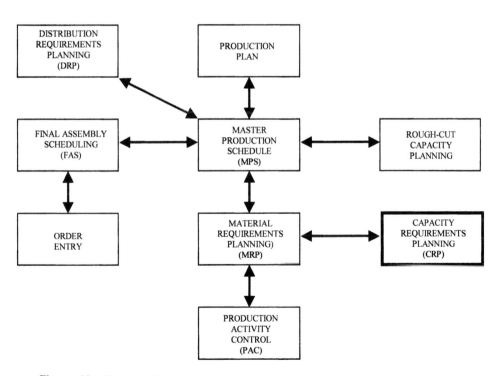

Figure 16 The closed loop system—capacity requirements planning.

Part: A	1	2	3	4	5	6
Planned Releases	0	10	10	0	10	0
LT = 2 Lot = 10						

Figure 17   The output of MRP.

Setting aside the issue that capacity requirements planning inherits the static time phasing created by material requirements planning, CRP suffers from the same problem as rough-cut capacity planning. Hours are dropped into buckets and aggregated against time that is not available.

1.   Looping Back to the MPS

The idea of looping back to the master schedule and correcting a capacity requirements problem at this point is useless. No real information has been collected to make that decision. However, even when this is attempted and a new master schedule is set the impact is very unpredictable. Since each operation in the routing is permanently connected by individual start and stop times, whenever a change is made to alter the load on one resource the change is also made on other resources. Unloading a resource that is loaded to greater than 100% may create problems on other resources.

In Fig. 20, orders for A and B create a load on resources R-1 through R-4 in period 9. Since resource R-2 appears to be loaded to 120% the master

Part No.	Oper.No.	Work Center	Que/Day	S/U Hrs.	Run Hrs.
A	010	Asmb	5	1	1.0
	020	Test	2	2	1.0
	030	Insp	3	1	1.0
B	010	Asmb	5	1	1.0
	020	Test	2	2	1.0
	030	Insp	3	1	.50

Figure 18   The route file.

Work Center	1	2	3	4	5	6
Assembly			11	11		11

Figure 19 Computing the load.

schedule must be leveled so that R-2's time is reduced to at least 100%. To accomplish this task order B is shifted in time so that it starts in period 10 instead of 9. Originally, the orders in period 10 did not create an overload problem. After the shift a problem was created when B's labor requirements were added to C (see Fig. 21).

Resource R-1 is now loaded to 120% and resource R-3 is now loaded to 167% in period 10. The problem is discovered only after running MPS, MRP, and CRP programs. So another iteration must be tried. The master schedule is adjusted and then the MPS, MRP, and CRP programs are run again.

The really interesting part about this process is that even if a master schedule is eventually set to a level load, the load indications are still bogus because of the problems already discussed. To this point the entire procedure has been an exercise in futility.

	R-1	R-2	R-3	R-4
Order A	10	20	10	10
Order B	10	20	10	10
Order C				
Total Hours	20	40	20	20
Load	80%	120%	85%	70%
	Period 9			

Figure 20 Unloading the MPS.

	R-1	R-2	R-3	R-4
Order A				
Order B	10	20	10	10
Order C	20	10	30	10
Total Hours	30	30	40	20
Load	120%	90%	167%	70%
Period 10				

Figure 21   Unloading the MPS.

IV. PRODUCTION ACTIVITY CONTROL

The "objective of the shop floor control system (SFC) is to execute the material plan" (*Manufacturing Planning and Control Systems*, 1988). It takes the output of material requirements planning and actually schedules the factory by setting the priority for orders in each work center. Also called production activity control (PAC), SFC represents the connection between the individual work centers on the shop floor and all that has occurred so far within the planning process (see Fig. 22).

A. Dispatching

Scheduling priority instructions are passed to individual workers or groups of workers via the dispatch report (see Fig. 23). Notice that work center R-1 has four work orders to be processed for week 24. Notice also that the total load is currently 77.50 hours and that the priority ratings are from .75 to 1.50.

   Though the dispatch report shown uses critical ratio to determine the priority of work, dispatch lists can use a number of different methodologies to create the priority arrangement of orders, including

- Slack time
- Slack time per operation
- Critical ratio
- Shortest operation next

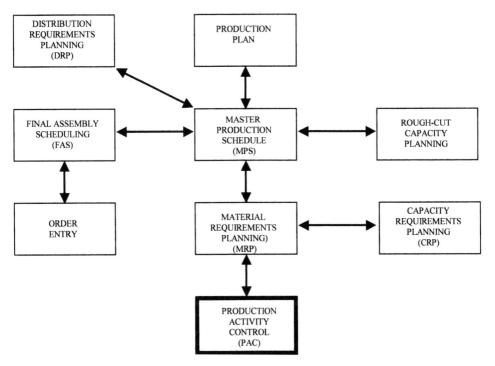

Figure 22 The closed loop system—production activity control.

Dispatch Report

Work Center: R-1   Desc. Mill   Week 24

Work Order	Part No.	Oper.	Description	Qty	Hours	Priority
W/O1234	PN456A	10	Mill	50	25.00	.75
W/O1235	PN456B	10	Mill	30	15.00	.98
W/O1236	PN456C	10	Mill	25	12.50	1.03
W/O1237	PN456D	10	Mill	50	25.00	1.50
				Total Load	77.50	

Figure 23 The dispatch report.

- First come, first serve
- Operation start time
- Due date
- Next queue
- Least setup

Each one of these methods has different focusing mechanisms and will create a specific outcome different than the next, such as reducing the number of jobs in queue or insuring that the next queue is full. However, the objective of shop floor control must be to "exploit the constraint and to subordinate the remaining resources to the way in which it has been decided to exploit the constraint." This is not what is intended by these methods, which thus serve to undermine the planning process.

- *Slack time* subtracts the sum of the remaining processing time from the time remaining. Time remaining is determined by subtracting the current date and time from the due date and time. Those orders which have the least slack are given the highest priority.
- The *critical ratio* method creates a ratio by determining the time remaining between the current date and the due date, then dividing this number by the lead time remaining. Obviously, the lower the number, the higher the priority.
- *Shortest operation next* concentrates on reducing the number of orders in the queue by taking those orders that can be finished the quickest.
- *First come, first serve* takes orders as they arrive and processes them regardless of need.
- The *operation start time* and *due date* scheduling methods attempt to minimize lateness by concentrating on the time that orders are to be shipped to the customer.
- *Next queue* looks for the shortest queue in the process after the current operation and feeds material to it so that downstream resources never run out of orders to process.
- *Least setup* looks at those orders with the least amount of time required to perform a setup operation. Similar in objective to the shortest operation next technique, least setup attempts to lower the overall number of orders in the queue by choosing those orders which have the least setup or where setup can be saved by processing two orders together with only one setup. The thought process is that the less time spent on setup and the more that is spent on production, the better the company will be financially. In addition, the assumption is made that reducing the number of orders at each work center will cause profits to increase.

These techniques can be categorized in two ways:

- Those which use local measurements to guide the decision of what orders to process next, such as shortest operation next, least setup and next queue
- Those which are based on due date criteria, such as critical ratio, operation start time, due date, and slack time

In maximizing local measurements it is hoped to maximize the global measurement of return on investment by reducing cost or minimizing queue time. Those sequencing techniques which use local measurements to guide the decision on what orders to process next have no basis in reality and from a global perspective can do more harm than good. They do not recognize that both parts and resources must be synchronized to maximize the generation of throughput.

As an example, next queue ignores the fact that the operation with the greatest queue may be the constraint and that the operation with the least queue may be a nonconstraint. Since the main issue of this technique is not to synchronize and protect those orders which will generate throughput, mismatched parts that must be assembled with those coming from the constraint will cause throughput to drop immediately.

Least setup suffers from the same problem as next queue and is a myopic way of looking at scheduling. If every work center used this kind of scheduling without interference in the way of expediting orders, the company would soon have no choice but to close its doors and declare itself insolvent. Reducing the total amount of orders at a work center rather than concentrating on exploitation or subordination means an overabundance of mismatched parts.

Those techniques which are based on a due date performance issue will probably be more effective in meeting the objective of satisfying the customer, but are not constructed from a complete understanding of how resources impact each other (*relationships between resources*) or what actions are necessary to accomplish the activities of exploitation or subordination. And yet, these are what must be done. For those techniques which use due date performance the question must be asked: to what date is the order to perform? There are a number of issues:

- In order to protect the schedule on the constraint the processing start and stop time of an order which is prior to arriving at the constraint must be to the constraint's schedule.
- In order to arrive at a proper start time for any operation there must be an aggregation of those things which can go wrong.
- In order to arrive at a proper start time there must be an understanding of what the current load is on the resource in question

and whether or not it threatens the constraint's ability to generate throughput.

- If the schedule for the order is at the constraint, then the sequencing of orders must be based on the concept of maximizing throughput at the constraint.

In any discussion of the sequencing techniques mentioned, no mention of how resources interface and the resulting impact on the schedule is ever given. In addition, there is an overwhelming assumption that no constraint exists or, that if it does, there will be no impact on how sequencing is to take place.

The objective of any scheduling policy must be to maximize the throughput generated by the constraint and to protect it. From a global perspective the techniques mentioned do not accomplish or help to accomplish this task.

B.   Statistical Fluctuation in Shop Floor Control

If each order within the schedule occupies a specific place in time and at the same time each resource fluctuates in its ability to deliver, the result is a situation where sometimes orders are delivered early and sometimes orders are delivered late. This is a naturally occurring phenomenon any time there is a schedule. Notice that in Fig. 24 the opportunity to be early is just as great as the opportunity to be late.

There is a naturally occurring bell-shaped curve. However, sometimes this curve moves to the right because of the impact other resources have when not delivered to this resource on time. Sometimes, even though the resource received its order early, it is not able to deliver on time due to something which

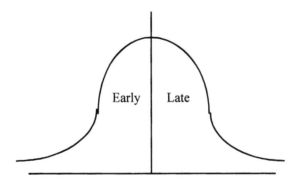

Figure 24   Statistical fluctuation in shop floor control.

occurred at this resource or because of its current load or both. These are naturally occurring statistical fluctuations and can be predicted as well as planned for. What cannot be planned for is human interdiction which prevents these events from occurring.

By distributing a dispatch report, this naturally occurring phenomenon can be circumvented rather than simply planned for. Unfortunately, it is human nature which consistently prevents orders from being early. By providing a start or due date, operators are not only being told when something is on time or late, but they are also told when something should not be started early. The problem here is that the impact of not starting a job early can only be detected in aggregate. In other words, unless it is known what the cumulative impact is across all resources, a decision cannot be made. In absence of this information it is better to be early whenever possible so that those resources which cannot perform, for whatever reason, are protected.

The issue is not that people will tend to slow down if they are early, although sometimes this is the case, but that nature abhors a vacuum and in the absence of sufficient motivation they will find what to them or their boss will be a higher priority. What is not realized is that being early from a global perspective can be very important. What is also not realized is that the current method of dispatching can be very detrimental (see Fig. 25).

Since each operation will fluctuate in delivery and the only side of the fluctuation that is allowed to occur is the late, the cumulative effect of not allowing resources to be early is that orders will be chronically late.

The real challenge is in finding out which resources need priority instructions and which do not. Certainly for those resources which have excess capacity, the dispatch list is detrimental to the goal of the company. For those which are loaded to 90% or better, it is probably mandatory.

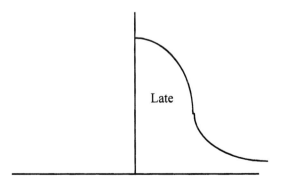

Figure 25 Statistical fluctuation in shop floor control.

C.　Dynamic Versus Static Data in Shop Floor Control

Lead time typically includes

- Setup time—the time required to prepare a resource for work.
- Run time—the time required to produce a particular part.
- Move time—the time spent after an operation moving the part to the next operation in the routing.
- Queue time—the time spent waiting at a work center before that work is performed.

Figure 26 represents a typical routing used in computing the lead time for a particular order. An order of ten part As at operation 10 will need 1 hour for setup, 10 hours for run time, 0.3 hours for move, and 5 days queue time for a total lead time of 6 days plus 3.3 hours, or approximately 6.5 days. Operation 20 will take 2 hours setup, 10 hours run, 0.3 hours move, and 2 days queue time for a total lead time of approximately 3.5 days. Operation 30 will need 1 hour for setup, 10 hours for run time, 0.3 hours move, and 3 days queue time for a total of approximately 4.5 days. The total lead time for part A is 14.5 days with queue time being almost 70% of the total lead time.

As seen, queue time is a very critical element in shop floor control. In most companies queue time makes up an average 80% of all lead time within the factory. Eliminated, it would mean that lead times in the factory would be 20% of the current figure. A lead time of 1 month would be less than a week, but how often is this number actually reviewed? Typically, queue time is maintained in the route file for each step in the operation. The question is how often are these data maintained. It is a full 80% of all lead time and is mostly ignored ... and for good reason. If a company has 4000 manufactured items

Part No.	Oper.No.	Work Center	S/U Hrs.	Run Hrs.	Move Hrs.	Q/Days.
A	010	Asmb	1	1.0	0.3	5.0
	020	Test	2	1.0	0.3	2.0
	030	Insp	1	1.0	0.3	3.0
B	010	Asmb	1	1.0	0.3	5.0
	020	Test	2	1.0	0.3	2.0
	030	Insp	1	.50	0.3	3.0

Figure 26　Typical routing for computing lead time.

with an average of five steps per routing, that's 20,000 individual record locations to manage so that 80% of your lead times are correct. This raises some key questions:

- How often is it monitored?
- If it is input for each part at each operation, how effectively can it be monitored?
- How often does it really change?
- Is there a better way?

The chances of these data being monitored effectively fall into two categories: slim and none. However, there are even greater problems. Since the true objective is to go from the gating operation to the creation of throughput the real issue is not the impact of those things that can go wrong "Murphy" on an individual operational basis, but the aggregate impact of Murphy across all operations. If queue time is placed between resources on an individual basis and then never allowed to change as the dynamics of the situation dictate, then it is at best a waste of time.

D. Finite Loading

Finite loading creates a detailed schedule of all orders and all work centers with stop and start dates assigned for each operation. The start and stop dates are computed using the output from MRP combined with lead times generated for each open or planned order. In addition, as the sequencing for each operation is generated, the amount of load on each resource is also determined. With infinite loading the placement of the load is unlimited by the amount of capacity available (see Fig. 27). With finite loading the placement of the load is limited by the amount of capacity available (see Fig. 28).

Notice that in the first period finite loading did not fill all the available capacity. However, beginning in the second period demand is pushed into succeeding periods (the future) so that it does not exceed capacity. Notice also that an additional period is used to absorb the demand as it is pushed out of period 6 into period 7. This method is known as finite set forward.

1. The Failure of Finite Loading

Finite loading presents some interesting problems. While it attempts to act in a rational manner by assuming that resources should not be loaded past their ability to produce, it fails in the attempt to solve the scheduling problem. Finite loading was originally created to use the output of MRP to generate a basic schedule and then augment it by not allowing resources

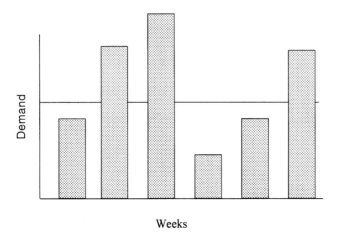

Figure 27 Infinite loading.

to be overloaded. At the same time it was to maximize resource utilization. This is fundamentally unsound for a number of reasons, including

- By pushing the load into later periods, it fails to protect the constraint in period 1.
- It is a victim of the static data used to generate the output of MRP.
- It also uses static data, including queue time between resources, from the route file during the scheduling process.
- It attempts to schedule all resources simultaneously.

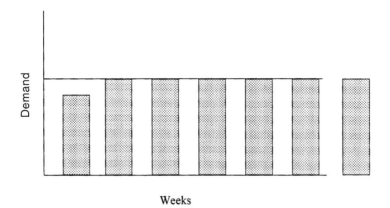

Figure 28 Finite loading.

If the creators of the finite loading process understood the impact of those laws which govern the way resources interface, they would have used as a prerequisite the first three steps of the five-step process of improvement. This is the model from which all systems must begin their development. Any system which does not use this process cannot successfully schedule the factory.

For a scheduling system to be effective it must be capable of

- Identifying the constraint
- Aiding in its exploitation
- Subordinating the remaining resources to the way in which it has been decided to exploit the constraint

These are fundamental and absolute. If a system is incapable of these three steps, it is incapable of successfully scheduling the factory. As seen in Chapter 4, if a system cannot produce a valid schedule, it cannot be used to make decisions.

In the discussion thus far, no portion of the system has been able to identify the constraint. Finite loading has not changed this issue. That is not its objective. Obviously, if the constraint has not been found, it cannot be exploited. But for discussion purposes suppose that the constraint has been identified. What is it about finite loading which aides in the exploitation process? It creates a schedule. However, to exploit means to squeeze the maximum out of the constraint. There is no evidence that finite loading uses any methodology to accomplish this task. If so, in any discussion of finite loading there would also be discussions of issues such as setup savings at the constraint.

Additionally, since finite loading schedules all resources simultaneously, it seems highly unlikely that a decision process needed to accomplish the exploitation and subordination phases could be done effectively. It is only after the exploitation phase is complete that any schedule can be created for the remaining resources. This means that a complete schedule is created by an iterative process. First the constraint is identified and exploited, then the remaining resources are subordinated. During the subordination process other resources may be found with a high probability of being chronically late in supplying the constraint. This is only possible after finding out what the constraint needs.

V. CONCLUSION

It is obvious at this point that the original assumptions upon which the traditional MRP II system was built were are and will continue to be erroneous. For the traditional system to fulfill its mission it must change. It must begin the migration process from data system to information system. However, apart from pointing toward some very basic issues such as the measurement system and the five steps of improvement, little has been accomplished so far other than to discredit some portions of MRP II. Building an information system based on reality is something else. How should it function? Upon what basis should it be built? At this point it is important to establish some basics that can be used to logically deduce what an information system for the manufacturing environment should do and how it should function.

VI. STUDY QUESTIONS

1. Define the content and the objective of the master production schedule.
2. Describe rough-cut capacity planning and explain why it fails in attaining it's objective.
3. Explain the input and output of the traditional MRP system.
4. Configure a gross-to-net requirements generation matrix. Explain how it works and why it fails to attain its objective.
5. Define three types of capacity that will have a major impact on how a factory should be scheduled.
6. What is the cascading effect and what is its impact on MRP?
7. What is the difference between a process and a transfer batch and what determines the optimal size for each?
8. Describe the premise behind the economic order quantity (EOQ) formula and list four assumptions which will invalidate it as a lot sizing alternative.
9. Define capacity requirements planning and explain how it works.
10. Give a brief explanation of the aggregation of demand and describe its impact on the CRP process.
11. What is meant by "looping back to the MPS" and what major flaw(s) is at work during the process?
12. Define *shop floor control* and the role of the dispatch report.
13. Name four priority methods commonly used in the traditional dispatching system, their objectives, and why they fail in meeting those objectives.
14. Describe the impact of statistical fluctuations on shop floor control and why dispatch lists can sometimes do more harm than good.
15. Describe the impact of static data on lead time in shop floor control.
16. What is finite loading and why does it fail to create an adequate schedule?

3
Synchronization Versus Optimization

The advent of new finite scheduling and supply chain optimization systems in recent years has served to underwrite one of the original problems identified in Chapter 1 of this book. It is very clear that before designing a system that is capable of synchronizing the factory or managing a supply chain, systems designers must understand the environment that these systems have been created to manage. In either of these environments synchronization between dependent resources is of monumental importance. A lack of synchronization will result in a less than desirable outcome.

It is equally important that systems managers tasked with systems selection understand the difference between those systems designed to optimize the factory or supply chain and those designed to promote synchronization.

I. CHAPTER OBJECTIVES

- To define the synchronization process and its benefits
- To define the optimization concept
- To demonstrate the negative impact of optimization

II. THE SYNCHRONIZATION CONCEPT

In order for money to enter the company through sales, a product must be delivered. In order for the product to be delivered, assuming more than one resource exists, the resources of the company must work together to accomplish this task. In turn, the customer places a regulatory condition on the system by requiring that the product be delivered on time. Other considerations must be given when the owner of the company decides that a profit must be made. This last statement means that there is a limit to how much inventory and operating expense will be tolerated to bring in a certain amount of throughput. Because of these limitations, synchronization becomes a priority of the system. To maximize throughput while minimizing inventory and operating expense, a high degree of synchronization is required. To a large degree, the higher the synchronization, the more effective the system is in supporting the goal of the company.

This concept poses some interesting questions:

- If it is required to maximize profit, how is synchronization created?
- How is inventory and operating expense minimized as a result?
- How does it support on-time delivery?

Note: The objective here is to isolate the synchronization process. A decision has been made to avoid issues such as the decision process. It is possible even when synchronization is at its peak to increase the amount of money entering the company by changing product mix.

A. High Versus Low Synchronization

A system is said to be synchronized when the resources contained within are working together to accomplish a common goal. However, this statement leads to very broad interpretation. A system may be synchronized, but to what extent? While synchronization is the goal of the traditional system, because of reasons described in Chapter 2, ERP must be considered loosely synchronized. So much so that it can cause major problems in helping the company reach its goal. What is needed is a higher degree of synchronization. This was the objective of the Kanban, or just-in-time, system created by Dr. Ono of Japan. However, an even higher degree of synchronization, and therefore profit, is available by understanding how synchronization is created and maintained.

The individual parts of the drive train of a car work together so that the output of the engine is transferred immediately to the transmission, the drive shaft, and finally to the wheels. There are two ways of looking at this analogy. One is that the drive train is one machine. The second way is to look at the drive train as made up of individual resources that feed each other and work so closely together that they are identified as one machine. The output of one resource is consumed immediately by the next; it delivers the right product to the right resource immediately and is considered highly synchronized. For a supply chain to be considered highly synchronized it must be designed as close to this model as possible.

B. Creating Synchronization

As discussed earlier, a necessary condition for any company comprised of a chain of events is to deliver to the customer what he wants when he wants it. This adds a two-dimensional quality to the definition required: the *what*, referring to a specific product, and the *when*, referring to a point in time. Since this is the activity that must occur before throughput enters the company it provides a sufficient starting point in determining what must be done.

While the diagram in Fig. 1 is very simple there are nuances about it that are not. Notice that a statement about preventing a shipment from being late or ensuring that it is on time permeates the entire diagram. This statement is designed to show that there is a need from one end of the resource chain to the other to focus efforts to ensure on-time delivery. It suggests that the right material will be available at downstream operations,

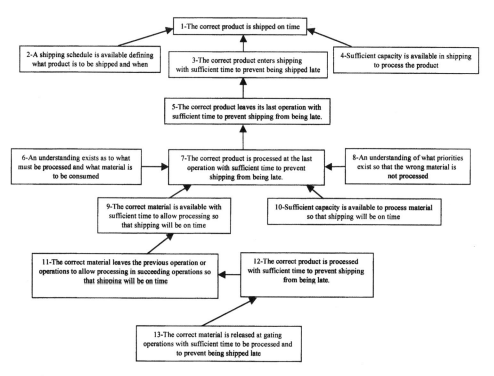

Figure 1   Diagram showing resource chain ensuring on-time delivery.

meaning that upstream resources must be working on a part that is to be consumed by downstream operations and in the proper order. It also suggests that material be released at the gating operation in the order in which it is to be processed and that the order reflects the desires and needs of the customer to be on time and have the proper quality. Additionally, to release material with sufficient time to be processed and not be shipped late means that an estimate of the amount of time necessary to produce the product and protect it from being late is known across all resources in aggregate. This last statement means that the laws involving probability and statistical fluctuation be applied. If the probability increases that a part will be late, the system can adjust.

1. Increasing the Degree of Synchronization

To increase synchronization between resources, as in the drive train analogy, parts that complete processing on one machine should proceed directly to the next and be processed without delay. This is accomplished by changing the transfer batch to a quantity of one. See Fig. 2.

Additionally, resources should not work on extraneous parts that will prevent immediate processing and delay shipping. The transmission does not operate in a forward mode while the drive shaft spins in reverse.

Note: Where the amount of synchronization may differ from the drive chain analogy is that some resources may not be available to take items for immediate processing. As an example, those resources whose capacity is limited may have a delay in processing. However, this does not mean that a transfer batch of one should not be applied. See Section II.A.5 on the impact of limited capacity on inventory later in this chapter.

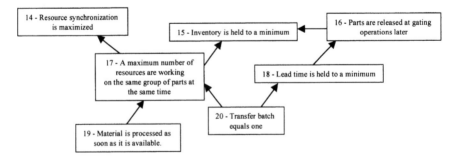

Figure 2 The impact of the transfer batch.

2. The Impact to Inventory of Using a Transfer Batch of One

The main benefit of using a transfer batch of one is that lead times as well as inventory are held to a minimum (see Fig. 3).

Referring to Fig., products A and B are being processed on resources 1 and 2. Each product is being processed left to right in lot sizes of seven. Each of the smaller blocks represents one part. The arrow indicates at what point a part is passed from resource 1 to resource 2.

Notice that product A is being passed to resource 2 after all parts have been processed on resource 1, while product B is passed after the first part is processed. During the processing of product B resources 1 and 2 have been highly synchronized due to the change in the transfer batch size.

Notice also that the timeline for B is half that of A. This means that parts can be released at gating operations later. If material is released later in time, it means that it does not need to arrive at receiving until later. Material that is not received is not inventory.

3. Estimating Release Times

Using a transfer batch of one does not guarantee on-time delivery. If the supply chain is synchronized but material is released at gating operations too late, the system would be unable to fulfill its goal of synchronizing to the market. In this case, the impact of the physical attributes of individual resources and the resource chain must be known so that an estimate can be made of the time required for material to flow from one end of the resource chain to the other. It also suggests that the estimation include the time used to offset for those things that will go wrong so that delays in shipping can be

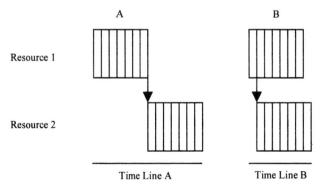

Figure 3 Minimizing lead time.

eliminated. This estimate would be used to establish the release time of material at gating operations. See Chapter 9 "Buffer Management."

4. Adjusting for Resource Limitations

Individual parts within the drive train have differing capacities. Run at an ever-increasing RPM, the pistons will leave the engine long before the drive shaft breaks. (This is why a red line is placed on the car's tachometer.) So the engine is regulated to the pace of the weak link in the chain of resources of the drive train. If a primary resource constraint exists within the supply chain, the resource must be synchronized to the market and the remaining resources synchronized to the constraint, including the release of material to the gating operation or operations that ultimately feed the constraint (see Fig. 4).

Notice that the order in which product is shipped, scheduled on the constraint, or released at the gating operation is exactly the same. The order of shipment is 10 product **A**s then 10 product **B**s. For purposes of synchronization, the constraint and gating operations perform in the exact same order.

A decision might be made to change the order in which the constraint is scheduled so that more products can be produced through setup savings. In this case, the release schedule and priorities for every other resource are modified to resynchronize with the constraint (see Fig. 5). The constraint, release, and shipping schedules now require the production of two orders of 10 **A**s, then two orders of 10 **B**s.

Release Schedule			Constraint Schedule		Shipping Schedule
10-A			10-A		10-A
10-B	Gating Operation		10-B	Shipping	10-B
10-A	○	CCR ○	10-A	○	10-A
10-B			10-B		10-B
10-A			10-A		10-A

Figure 4 Constraint synchronization.

Release Schedule		Constraint Schedule		Shipping Schedule
10-A		10-A		10-A
10-A	Gating Operation	10-A	Shipping	10-A
10-B		10-B		10-B
10-B		10-B		10-B
10-A		10-A		10-A

Beneath the headers: Gating Operation (circle) under Release column, CCR (circle) under Constraint column, Shipping (circle) under Shipping column.

Figure 5   Schedule resynchronization.

Note: Great care must be used to ensure that when the constraint schedule is created the resulting impact on the shipping schedule will not have a detrimental impact for on-time deliveries to the customer.

5.   The Impact of Limited Capacity on Inventory

Inventory will tend to collect at those resources having a limited ability to keep up with the demands of the rest of the supply chain. As seen in Fig. 6, inventory is collecting in front of the constraint.

6.   The Impact to Predictability

Because of the increased synchronization of the supply chain the predictability of the schedule increases. As a result, overtime and expediting will go

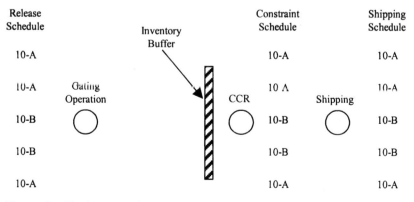

Figure 6   The impact to inventory.

down, while on-time deliveries will increase. A decrease in overtime results in a decrease in operating expense.

III. OPTIMIZATION SYSTEMS

Supply chain optimization systems are designed to sub-optimize between two or more measurements, the objective being to maximize profits using local measurements. By controlling specific variables, usually associated with cost, the user hopes to limit costs and thereby maximize profit.

Typically, optimization systems will use an array of complex mathematical formulas to manage the sub-optimization process. The user becomes enamored with the concept that somehow these formulas will help him reach his goal. Since this is probably in line with his current way of thinking about profitability, the user will see nothing wrong. Of course, nothing could be further from the truth as seen in the following example:

Figure 7 represents a supply chain with three manufacturing plants and two warehouses. Manufacturing plants 1 and 2 create their products and deliver them to their respective warehouses. The warehouses then deliver to manufacturing plant 3. Manufacturing plant 3 assembles the parts then

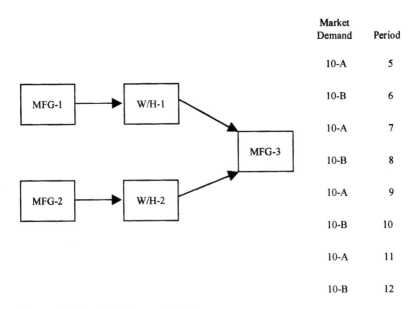

Market Demand	Period
10-A	5
10-B	6
10-A	7
10-B	8
10-A	9
10-B	10
10-A	11
10-B	12

Figure 7 Supply chain optimization.

delivers to the customer the finished product. From this diagram it is easy to see that if the proper amounts of parts are not produced at plants 1 and 2 and forwarded on time and at the right amounts, then MFG-3 will be unable to deliver to the market on time.

At this point, the question becomes, what actions can cause these plants to not deliver on time? Two basic models have been chosen:

- The optimization of the lot size during manufacture
- The optimization of transportation costs within the chain

A. Optimization of Lot Size During Manufacture

The optimization of the lot size during manufacture refers to, but is not limited to, the economic order quantity. As seen in Chapter 2, the economic order quantity sub-optimizes between setup and carrying cost to find the optimal lot size. While the assumptions behind the economic order quantity algorithm can be invalidated in a number of different ways, what is being emphasized is the impact on the supply chain by the lot size chosen.

As mentioned, in order for MFG-3 to produce to the market schedule it must receive parts from the preceding facilities in the proper amounts and at the proper time. The following assembly diagrams will help to determine part requirements.

Part **A** contains one A1 and one A2; part **B** contains one B1 and one B2 (Fig. 8). A1s and B1s are made and stored in MFG-1 and W/H-1, respectively, while A2s and B2s are made and stored in MFG-2 and W/H-2.

Assuming a lead time of two periods the market demand from MFG-3 would produce completion schedules for MFG-1 and MFG-2 as in Fig. 9. Notice that MFG-1 and MFG-2 have been synchronized to MFG-3's schedule.

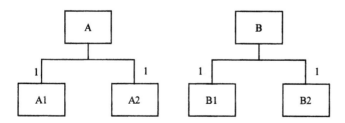

Figure 8 Parts structure.

Periods

	3	4	5	6	7	8	9	10
MFG-1	A1	B1	A1	B1	A1	B1	A1	B1
MFG-2	A2	B2	A2	B2	A2	B2	A2	B2

Figure 9 Schedule completion requirements.

Note: For the sake of brevity, an assumption is being made during optimization that the setup costs at MFG-1 are high resulting in a combination of parts demand covering more than one period. (Review economic order quantity in Chapter 2.)

Figure 10 reflects the schedule after optimization. The schedules have now been desynchronized and MFG-3 does not have the right combination of parts to meet its schedule. Of course, the overall impact on the company's profitability is negative. The amount of money entering the company through sales will be reduced due to the lack of synchronization caused by the optimization at MFG-1, while operating expenses will remain the same. Total inventory in the system will go up due to MFG-3s inability meet market demand. Interestingly, the plant manager at MFG-1 will probably meet his efficiency numbers.

B.   Optimization of Transportation Costs

The optimization of transportation costs usually occurs to reduce the costs associated with transferring items. Transferring can occur anywhere in the supply chain. However, in this case reference is being made to those costs associated with transfers that occur between facilities in the supply chain. As an example, after products have been produced at MFG-1 they must be transferred to warehouse W/H-1 and then to the final assembly facility

Periods

	3	4	5	6	7	8	9	10
MFG-1	A1	A1	B1	B1	A1	A1	B1	B1
MFG-2	A2	B2	A2	B2	A2	B2	A2	B2

Figure 10 Optimized completion schedule.

Product Cost		Net Profit	
Material	600	Sales Price	3500
Labor	600	Total Cost/Unit	2550
Overhead	1200	Net Profit	950
Transportation	150		
Total Cost/Unit	2550		

Profit = 27%

Figure 11   Transportation costs for products A1 and B1.

MFG-3. Costs associated with transportation for products A1 and B1 are shown in Fig. 11.

In the traditional sense, the $2550 is the cost of raw material for MFG-3. To transfer both parts A1 and B1 from MFG-1 to MFG-3 costs $150. However, by combining trips so that they cover more than one period, transportation costs can be reduced by 50%, or $75. The company's net profit will supposedly increase to 29% per part (see Fig. 12).

Note: No attempt has been made to justify the charges such as whether the company owns the trucks being used, etc. Only the impact of the optimization of transportation costs on synchronization is being reviewed at this time. Whether or not the cost reduction results in a reduction of operating expense is a different issue.

Figure 13 shows the impact of the decision on the shipping schedule. A delay has taken place in the delivery of A1 to MFG-3. The delay results in an increase in lead time. What normally takes two periods to deliver now takes three, thereby making it more difficult to change production if the customer changes his mind. Customer service will be reduced.

Product Cost		Net Profit	
Material	600	Sales Price	3500
Labor	600	Total Cost/Unit	2475
Overhead	1200	Net Profit	1075
Transportation	75		
Total Cost/Unit	2475		

Profit = 29%

Figure 12   Transportation costs optimization.

Periods

	1	2	3	4	5	6	7	8
MFG-1		A1		A1		A1		A1
MFG-1		B1		B1		B1		B1
MFG-2	A2	B2	A2	B2	A2	B2	A2	B2

Figure 13   Impact of optimizing transportation.

Because lead time has increased, the amount of inventory in the system will go up. In order to meet demand, parts will be released into production earlier.

Since more parts will be waiting to be transferred, production problems will go unnoticed for longer periods of time causing quality to decline.

Parts being delivered from MFG-2 must wait an additional period before being used at MFG-3 because of a lack of matching parts from MFG-1. Because of this the schedule at MFG-2 must be changed as well.

If the additional lead time required to create and deliver to the market pushes the total lead time past the customer tolerance time, additional finished goods items must be stored and a forecast created. Errors in the forecast will increase finished goods inventory due to a lack of sales for that particular product.

C. Probability

Optimization systems typically do not understand the issues regarding probability. Their goal is usually to optimize between inventory, labor, and on-time delivery. However, in order for synchronization to occur probability must be considered. As seen in Chapter 2, lead times are dynamic; they increase and decrease based on the load on the system. The reason that lead times increase or decrease is that the probability of on-time delivery increases and decreases as resource loads change: the higher the load, the less the probability that a part will be processed on time.

D. Exploitation

An additional problem exhibited in the typical optimization system is that no attempt is made to optimize/exploit the amount of material being produced at the constraint. In fact, the algorithms used for the optimization of inventory, labor, and on-time delivery make it virtually impossible to exploit the constraint. Each scheduling technique excludes the use of the other.

E. Buffering

In order for synchronization to occur, buffers must be placed in strategic positions throughout the schedule and allowed to increase or decrease as probabilities change. These buffers are used to offset for those things that can and will go wrong (see Chapter 5). Since things will go wrong that will destroy synchronization, correct buffering is mandatory. The concept of buffering seems to be totally alien to the typical optimization system.

IV. CONCLUSION

Assuming the constraint is at an internal resource, synchronization of either the factory or the entire supply chain maximizes the ability of a company to make more money enter the company while minimizing inventory and operating expense. Even when the constraint is external, maximizing synchronization provides distinct advantages in managing inventory and operating expense so that they are kept to a minimum while providing the ability to exploit the external (market) constraint by ensuring a high degree of on-time delivery. Additionally, synchronization has the ability to release excess capacity to aid in the exploitation of the market by providing increased pricing flexibility.

Optimization systems do not take into consideration the need to maintain or maximize synchronization. They simply optimize local measurements. However, it is this optimization process which destroys synchronization and thereby reduces the overall profitability of the company.

V. STUDY QUESTIONS

1. Define the synchronization concept and explain why it is important.
2. Define how synchronization is created and maintained in a manufacturing environment.
3. Explain the difference between a high and low degree of system synchronization.

4. What impact does buffering have on synchronization and what are the key points within the schedule where buffering must occur to support synchronization?
5. Define the typical supply chain optimization system and why it destroys synchronization.
6. Explain the impact of the optimization of lot size during manufacture on the synchronization process.
7. Explain the impact of the optimization of transportation costs on the synchronization process.

4
Establishing the Characteristics of an Information System

What should a manufacturing information system do? This is not a trivial question. Stating that it should schedule the factory, compute material requirements, or create a master schedule is not enough to answer the question. The system may need to accomplish these tasks, but if these are the criteria by which the system is designed, it will probably be subject to major flaws. To really understand what an information system should do requires careful consideration.

I. CHAPTER OBJECTIVES

- To create an understanding of the characteristics of an information system
- To create a relationship between the information system and the decision process
- To understand the organization of the system

II. DISTINGUISHING BETWEEN DATA AND INFORMATION

In order to build an information system there must be a distinction between what is called *data* and what is called *information*. As mentioned in Chapter 1, information is the answer to a specific question asked, while data are everything else in the system. This is a very important clue. It means that in order to create an information system there must be an understanding of the decision process to be used. This can serve to be a major stumbling block for most developers. The decision process has proven to be very much different than what has been taught for the last 50 to 100 years. However, in building an information system the developer must be able to make this distinction. There must be knowledge of what questions to ask and how decisions are made. The first step in understanding how decisions should be made is to determine the goal of the system.

III. DEFINING THE GOAL

To be effective, an information system must help a company attain its goal. Obviously, it is important to have a definition of what the company goal should be. For the purposes of this book an assumption will be made that the goal of the company will be to "make more money now as well as in the future" (*The Haystack Syndrome*, 1990).

   Since this statement has had major impact on what an information system should do and is actually the basis for the system's creation, it's very important to understand what it really means. If the objective of the company is simply to make money and a system is needed to help in the process, the definition of the system would be much different than a system designed to help a company make more money now as well as in the future, the key words being *more* and *in the future*. For a company to *make money now* might mean that a system would be needed to help sell all the assets of the company. Once accomplished, the company would not be capable of creating additional money in the future. Making *more money* means that the company wishes to make more money than it is currently capable of making. A system to accomplish this task will be different than that required to maintain the current flow.

IV. ESTABLISHING THE APPROACH

The question that must be asked at this point is what must a company do to accomplish its goal? Answering this question will help considerably to further define the characteristics of the system. The physical laws of the company involve what is called the *dependent variable environment*. The dependent variable environment refers to the environment that has multiple resources depending on each other, and each resource fluctuates in its ability to perform. Since each company is made of chains of events that occur as a result of the interaction of multiple resources, the laws involving dependency and statistical fluctuation dictate how they will interact.

   The only valid method of improvement for this environment is the five-step process of improvement defined in *The Theory of Constraints: Applications in Quality and Manufacturing* (1997). If this is the method by which a company intends to improve and an information system is desired to help in this effort, then it stands to reason that an information system designed to help the company attain its goal should be based on the five-step process:

1. Identify the constraint.
2. Exploit the constraint.
3. Subordinate the remaining resources.
4. Elevate the constraint.
5. Repeat the process.

From a design perspective this is actually very good news. This means that the creation of the information system can be accomplished iteratively. The first step is creating the ability to identify the constraint. The second step is to exploit it. And so on. If each step can be accomplished successfully without creating new logistical or policy constraints, then the development of the information system will be successful.

V. EXPANDING THE DEFINITION

Because the information system must deal with the physical attributes of the environment in which it must work, the previously stated goal requires expanding. To be effective the system must help the company make more money now as well as in the future, *given the current situation.*

This addition to the statement adds another dimension. If a system is to be based on the current situation, it stands to reason that the current situation be known. This is a major clue to another of the characteristics of the system. *It should resemble the environment in which it must perform.* To do this it should be capable of recreating the physical laws that govern the environment in question. To ensure that the system can perform its function means that it should *not* contain within its design any logistical or policy constraints. This is why it is so important to eliminate those characteristics in the traditional MRP II system.

Note: companies that operate on a different principle such as a not-for profit basis may have a different goal and therefore may have a different definition of what an information system should be.

VI. THE INFORMATION SYSTEM AND THE DECISION PROCESS

Many of a company's problems (90–99%) in attempting to implement the five steps will surface in the form of policy constraints. In other words, for whatever reason and no matter at what level of the company, a decision will be made not to make more money. This sounds rather amazing until it is considered that most people are still working with the inertia created by cost accounting and are usually unaware that they have made a decision or taken a specific action that will hurt their company. In fact, even when presented with clearcut evidence to the contrary, people will insist that they were correct. This is one of the main reasons why changing a company can be so difficult. However, if information is the answer to a specific question, then it is not until after a close examination of the decision process is made that an information system can be built. The following examples demonstrate some of the issues.

A. Making a Sales Decision: Accepting an Order

The traditional decision process uses the difference between standard cost and the sales price (see Fig. 1). If the sales price does not exceed standard cost by a given percentage, the salesman turns down the offer. However, as seen in Chapter 5 of *The Theory of Constraints: Applications in Quality and Manufacturing*, what the salesman really needs to know to make the decision is much different. It includes

- The lowest throughput per unit of the constraint that is currently being produced
- How much throughput per unit of the constraint will be generated by the new order
- Whether the new order will use excess or protective capacity by producing the product internally
- Whether accepting the new order will create a new constraint

Note: Throughput per unit of the constraint is computed as the sales price minus raw material divided by the total constraint time for the product.

| | | | Standard |
Material	Labor	Overhead	Cost
100 +	50 +	100 =	250

| | Sales | Standard | |
	Price	Cost	Margin
	350 -	250 =	100

Figure 1 Making a sales decision.

In order to help the salesman make a decision the system must determine

- The constraint
- The amount of time each product spends on the constraint
- The cost of raw material for the product being produced
- The impact of using excess capacity
- The impact of using protective capacity

Since a key ingredient in implementing the decision process has been in understanding the impact of the constraints of the system and the impact of protective or excess capacity, to fulfill these requirements demands that a valid schedule be produced which will be capable of determining, for instance, where the constraint is.

B. Making a Quality Decision: Scrap Versus Rework

Another type of decision needing an answer is scrap versus rework. Would it be better from a global perspective to scrap a particular product or rework it?

In Fig. 2, a sales order for 100 product As each sold for $100 has a raw material content of $25 and is processed on the constraint. After processing 50% of the parts, ten are found to be defective. Reworking the parts will use an additional 10 minutes of constraint time and $5 worth of raw material. The question is should these parts be scrapped or reworked and what information is needed to make the decision.

PRODUCT	CCR TIME/UNIT	CCR TIME AVAIL
A	15	1500 MIN

| ORDER = 100 | SP = $100 | RM. − $25 |

DEFECTIVE PARTS	TIME/UNIT	MATERIAL
10	10	$5

Figure 2 The quality decision—sample data.

If the parts are scrapped, then of course they must be replaced. The original parts plus the remaining 50 must be produced creating $4500 worth of throughput and using 900 minutes of constraint time. The scrapped parts will require additional raw material totaling $250 that must be subtracted from the total throughput generated. Throughput per unit of the constraint is therefore $4250 divided by 900 minutes, or $4.72 per minute earned from the constraint (see Fig. 3).

Figure 4 shows the result if the ten product As are reworked. Ten As are processed on the constraint for 10 minutes using 100 minutes of constraint time and $5 additional raw material. Throughput generated is $700. The remaining 50 pieces are processed using 750 minutes and creating $3750 in throughput. Total throughput generated is $4450 and the total time used on the constraint is 850 minutes. Throughput per unit of the constraint is $5.23.

Since the throughput per unit of the constraint is higher for rework, the parts should be reworked.

What was needed to make this decision?

- The total amount of throughput that was to be generated by either scrap or rework (throughput = sales price – raw material)
- The amount of time to be used on the constraint for either scrap or rework

To help the foreman the system must be able to determine

- The constraint
- The amount of time each product spends on the constraint

			(T)	CCR TIME
THROUGHPUT	=	60 (100 - 25)	= $4,500	900
MATERIAL	=	10 X 25	= $ 250	
			$4,250	900

$$T/uc = \frac{\$4,250}{900} = \$4.72$$

Figure 3 The quality decision—throughput per unit of the constraint if parts are scrapped.

			(T)	CCR TIME
THROUGHPUT	=	10 (100 - 30)	= $700	100
THROUGHPUT	=	50 (100 - 25)	= $3,750	750
			$4,450	850

$$T/uc = \frac{\$4,450}{850} = \$5.23$$

Figure 4 The quality decision—throughput per unit of the constraint if parts are reworked.

- The cost of raw material for the product being produced
- The impact of using excess capacity
- The impact of using protective capacity

VII. ESTABLISHING THE MEASUREMENTS

The effectiveness of any information system must be judged in the context of how well the system is able to aid the company in performing relative to its goal. The global measurement of return on investment provides adequate proof that the system is accomplishing its primary function. However, it does not provide a method of focusing improvements so that the company understands what is required to reach its goal. In other words, if a decision is to be made at a local level to improve the performance of the company, return on investment will help to determine the impact of the decision, but it will not help determine what to improve. To do this will require that local measurements be established. This was the original objective of cost accounting, but those systems that have relied on cost accounting to make decisions have been greatly disparaged ... and for good reason.

> Not only has the effectiveness of cost accounting and activity-based accounting been disproved, but also the procedures have been determined to be the root cause of major problems leading to poor corporate performance.

The traditional measurements of effectiveness used at the local level can actually prevent the generation of valid decisions.

In order to understand what will be necessary to improve the company relative to its goal, local measurements must be employed which can be readily predictable from a global perspective. The measurements of choice are

- Throughput—the rate the system generates money through sales.
- Inventory—all the money invested in purchasing the things the system intends to sell.
- Operating expense—all the money the system spends in turning *inventory* into *throughput*.

Dr. Eli Goldratt presented these measurements in his book *The Race* (1986) and, as seen in *The Haystack Syndrome*, is an important ingredient in defining how an information system should work. The key issue when answering a question for a company whose goal is to "make more money now as well as in the future" is what will be the impact on throughput, inventory, and operating expense given a specific action? To create an information system these questions must be answered. In addition, the system must be capable of focusing efforts in proper order. Since "inventory and operating expense exist to produce and protect the throughput figure, understanding more about what is to be done to improve throughput is a prerequisite to dealing with the other two measurements" (*The Theory of Constraints: Applications in Quality and Manufacturing*).

VIII. THE ORGANIZATION OF THE SYSTEM

It should be clear by now that in order to simulate the decision process, or to answer the "what if" questions, there must be a *valid schedule*. However, it is not enough just to create a schedule. The schedule must be protected from those things that can go wrong. This protection must take into consideration where the protection is needed and be dynamic in nature so that it can change as the situation dictates. It must be capable of *controlling* the activity of the nonconstraints so that they are subordinated to the constraint. So the organization of the system should be as follows:

- Schedule—identifies the system's constraints, exploits them, and subordinates the activity of the nonconstraints.
- Control—compensates for those things which will go wrong.
- What if—replaces cost accounting in answering questions.

IX. SUMMARY

Since the goal of the company plays a major roll in what an information system should do, the real issue becomes what the goal of the company is and what the physical laws dictate must be done in order to improve.

An effective information system designed for this environment should

- Aid in the implementation of the five steps of improvement
- Resemble the environment by recreating the physical laws which govern it
- Use a measurement system which is based on reality
- Be void of any policy constraints

X. STUDY QUESTIONS

1. Define the goal of an information system.
2. Describe the connection between the goal of the information system and its basic structure.
3. What is the distinction between what is called *data* and what is called *information*?
4. What is the connection between the information system and the decision process?
5. What is the formula for computing throughput per unit of the constraint and why is it important?
6. What must a salesman know to make a determination of whether to accept an order and how does this differ from the traditional view?
7. List and explain the three absolute measurements defined by the theory of constraints.
8. Explain why these three measurements are important.
9. What is the dependent variable environment and what is its impact on the information system?

5
Defining the Structure

This chapter presents the structure of the system from the global perspective and defines the individual relationships of each segment, including input and output requirements.

I. CHAPTER OBJECTIVES

- To provide a global picture of how the system is to be structured
- To define the input and output relationships

II. THE THEORY OF CONSTRAINTS–BASED SYSTEM

Figure 1 is a diagram of the theory of constraints–based system designed to implement the five steps of improvement. Notice that the master schedule still drives the input to the system. It is a logical place from which to combine forecast and sales order data, show available to promise (ATP) for future sales requirements, and relate the status of orders.

However, unlike MRP II, rather than serving as the input to material requirements planning, the output of the master production schedule (MPS) will be the input to the process designed to identify the primary constraint. To properly schedule the factory the new system must be capacity sensitive, and since a valid schedule must be generated based on the system's constraints, the first step is to identify the primary constraint. It is during this process that material is allocated so that requirements will not be generated for material already in the system. The demand for additional raw material and its timing can only be determined after all the constraints have been identified and a schedule created for each. It is only after the needs of the constraints are

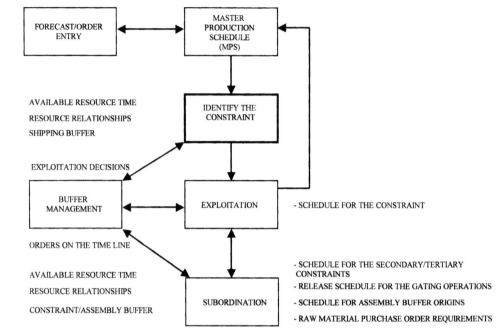

Figure 1   The TOC-based system.

known that a release schedule can be produced from which to glean the amount of raw material required.

The next step is to use the output of the identification phase to direct exploitation. It is during this phase that the schedule is created for the primary constraint and decisions are made for how the maximal amount of throughput is to be generated. Notice that from the exploitation phase there is a feedback mechanism to update the master schedule. After a firm schedule for the primary constraint has been set, if their are any changes that will require modification of the master schedule, then the new dates can be passed.

Once the schedule has been set, it is important to ensure that not only the remaining resources deliver to the constraint what it needs, but also that the remaining resources do not interfere with the constraint's ability to deliver on time. The subordination phase takes the output of exploitation and subordinates the activities of the remaining resources to the schedule created for the primary constraint. During this process the system attempts to eliminate any conflicts between the constraint and all other resources (see Chapter 8). If the conflicts cannot be eliminated automatically, then the user is given the opportunity to intercede manually. When this fails, conflicting orders on the primary constraint are pushed into the future along with the respective due date.

Buffer management is a methodology for monitoring and controlling the impact of those things that can go wrong (see Chapter 9). As changes occur in the system's ability to respond effectively, the buffer size used as input to the identification, exploitation and subordination phases can be increased or decreased to gain or lose protection.

III. DEFINING THE INPUT TO THE SYSTEM

A. Identification

The input to the identification phase consists of

- The master production schedule
- Resource relationships
- Available resource capacity
- The shipping buffer

The initial input of the TOC-based master production schedule is not representative of the traditional MPS. While it consists of a list of end items with relative quantities and due dates, no netting or consolidation of MPS demand into time buckets occurs at this time. It usually does *not* represent "what can and will be produced." To this point, no attempt is made to determine any capacity requirements. The MPS is simply a wish list until direct feedback has been received from the final version of the schedule for the primary constraint.

Resource relationships are defined in what is termed the *net*. The net, a creation of the Goldratt Institute, includes the product flow diagrams for all products in the master schedule and is used for computing material requirements, determining which resources are used to produce what products, and understanding how resources interface.

Obviously, if a resource limitation is to be identified, the resource availability in time increments must be involved.

The shipping buffer is a time mechanism for protecting the delivery of products to the constraint and is used for the initial placement of orders and resource demand on the timeline or schedule horizon.

B. Exploitation

The input to the exploitation phase includes

- The sequencing of orders on the timeline

- Decisions of how to exploit the constraint
- Setup and run times for each operation
- Available resource capacity

During the identification process, orders are placed on a timeline that runs from time zero, or today, to the end of the schedule horizon. After initial placement, the orders are leveled by pushing the demand toward time zero (see Chapter 6). It is the output of this process that is the input to exploitation.

During the exploitation phase, a schedule is created for the constraint. However, by definition, the constraint is a resource that is limited in capacity. If the constraint is to be exploited, decisions must be made to maximize the amount of throughput that can be generated. It is at this time that the system considers issues such as overtime, setup savings, and off-loading to maximize resource availability.

While the system can automatically apply or make recommendations of where overtime should be spent and what orders should be combined to save productive capacity, it is the user who must finally decide.

During the exploitation phase, setup and run times are used to determine start and stop times for each order across the constraint and to aid in simulating the decision process by performing tasks such as setup savings.

As in the identification phase, in order to exploit the constraint, the availability of the capacity to be applied to each order should be known.

C.   Subordination

The input to the subordination phase includes

- The schedule of the primary constraint (exploitation)
- The relationships between resources
- Available resource capacity
- Constraint/assembly buffers
- Setup and run times

As in identification and exploitation, the relationships between resources, the available resource time, and the setup and run times for each order must be known. However, since the objective of subordination is to deliver to the constraint what it needs, input to subordination must include the demands of the constraint. This means that for those part/operations processed prior to arriving at the constraint, the schedule created during exploitation must be used as input along with the requirements of the master schedule. In addition, the structure must include a method for protecting the constraint. *Constraint buffers* are used to aggregate the impact of those things which can go wrong on those resources which feed the constraint. *Assembly buffers* aggregate the impact of those things which can go wrong on those legs which are not constrained but must deliver parts to be combined in assembly operations with parts from the constraint.

IV. DEFINING THE OUTPUT OF THE SYSTEM

A. Identification

The output of the identification phase is very simple. It must feed the exploitation phase with

- The identity of the constraint
- Orders on a timeline

Without doubt, before the constraint can be exploited it must be identified. During the identification process, orders are placed on the timeline for the constraint.

Note: The timeline for the constraint is the scheduling horizon.

B. Exploitation

As discussed the output of the exploitation phase is a schedule for the primary constraint. The schedule includes start and stop times for every order within the planning horizon.

C. Subordination

The output of subordination includes

- The schedule of the secondary, tertiary, and remaining constraints
- The schedule for assembly buffer origins
- The release schedule for the gating operations
- Raw material purchasing requirements

Once the primary constraint has been identified and scheduled those resources that may be unable to deliver to the constraint what it needs due to a lack of adequate protective capacity must themselves be exploited to ensure their ability to deliver to the constraint. Secondary or even tertiary schedules may be produced to accomplish this task.

Note: A key issue in subordinating one schedule to another is to ensure that there are no conflicts between the two.

It is very important that a schedule be set for assembly operations having one leg coming from the constraint and another coming from a nonconstraint. The objective is to prevent the possibility of mismatched parts arriving from nonconstraining legs and to ensure that once parts have been created on the constraint that they do not wait for parts coming from the nonconstraining leg. To perform this function the assembly schedule is used. The assembly schedule coordinates the activity of the assembly buffer origin and determines the release schedule for raw material at the gating operations.

The release of raw material to the gating operations is planned by the subordination phase to ensure the proper sequencing of orders throughout the plant.

Once the releases of orders to the gating operations have been planned, material purchases can be made. The launching of raw material orders is planned using the lead time data stored in the item master for each raw material part plus the release schedule for the gating operations.

V.   ESTABLISHING THE RELATIONSHIPS
BETWEEN RESOURCES

As mentioned earlier relationships between resources for the schedule are established by the creation of what is termed the *net*. The net is a recreation of the way in which products flow through the factory and includes

- The combination of all product flow diagrams
- Work-in-process inventory
- Operational setup and run times for which demand appears in the master schedule

The product flow diagram is a combination of individual part/operations and resources on which the product is to be processed. The part/operation is the smallest building block within the net and is a combination of the bill of material and route files. In one character string the part/operation can identify what part is to be processed at a specific routing step (see Fig. 2). From the bill of material and route file shown, a combination of part A with operation 10 would create part/operation A/10.

Part/operations are placed in sequence along with resource information to form the product flow diagram The product flow diagram defines how the completed part is to be produced (see Fig. 3). Notice that purchased part C feeds part/operation B/10 on resource R-6. B/10 feeds B/20 on resource R-5. This process continues until the final product A is complete at A/30 on resource R-1.

Figure 4 represents the net. Notice that the flow of the net is the same as the product flow diagram. Raw material part C is received into resource R-6 to perform part/operation B/10. Resource R-6 then feeds resource R-5

INDENTED BILL OF MATERIAL			ROUTING			
			PART	OP.	S/U	RUN
A		MAKE	A	10	10	15
				20	15	5
	B	MAKE		30	10	10
		C PURCH.	B	10	10	5
				20	15	10
	D	MAKE		30	10	10
		E PURCH.	D	10	10	15
				20	15	15
				30	10	10

Figure 2 Creating the part/operation.

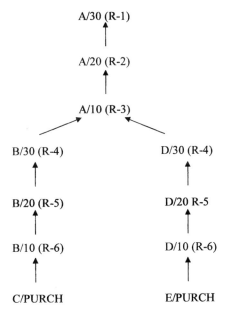

A/30 (R-1)

A/20 (R-2)

A/10 (R-3)

B/30 (R-4)   D/30 (R-4)

B/20 (R-5)   D/20 R-5

B/10 (R-6)   D/10 (R-6)

C/PURCH   E/PURCH

Figure 3 The product flow diagram.

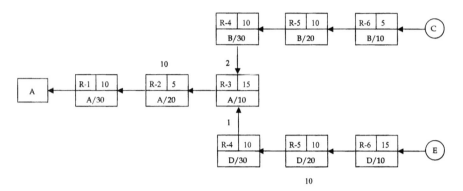

Figure 4   The net.

through the connection of B/10 to B/20. Additional data within the net include resource run time, bill of material quantity relationships, and inventory. Part/operation B/10 is processed on resource R-6 for 5 minutes. There are currently ten parts which have been completed by D/20 and are ready to be consumed by part/operation D/30. Part/operation A/10 is created by assembling two B/30 parts with one D/30 part at resource R-3. (Setup time was not included due to a lack of space.)

As discussed the net is the major source of data used to

- Identify the constraint.
- Generate the schedule (exploitation).
- Subordinate the remaining resources.

The net provides two major advantages:

- A very large number of data files can be reduced and stored in a very small amount of space within the system.
- The time needed for creating a schedule or determining material requirements can be greatly reduced.

Since the amount of space used is very small in comparison to the vast number of files currently being used, the entire net can be maintained in resident memory. By reducing the amount of I/Os needed for processing, the speed of the system can be greatly enhanced.

VI.   SUMMARY

The TOC-based information system is very much different from the traditional MRP II system. The basic structure and processes have been greatly modified to include the goal of the corporation as well as the physical laws which govern the manufacturing environment. Additionally, new innovations, like the net, have also been designed to take advantage of new computer technologies and enhance the speed at which information can be obtained.

VII. STUDY QUESTIONS

1. Prepare a schematic of the TOC-based system and define the input and output requirements of each segment.
2. Define the term *part/operation*.
3. Define what is meant by a product flow diagram and explain its primary function.
4. Create a bill of material and routing for one product produced in a typical job shop environment and then prepare a product flow diagram presenting the flow of part/operations involved.
5. What is the net and how is it constructed?
6. What advantage(s) does the net provide in constructing an information system?
7. Define the terms *identification*, *exploitation*, and *subordination* in reference to the TOC-based manufacturing system and describe their basic functions.
8. Describe the function of buffers. List three different types in relation to the TOC-based manufacturing system and explain how each is used.
9. What is meant by the term *timeline*?
10. In the TOC-based system, how is the master production schedule similar to the way in which it is used in the traditional MRP system?

6

Identifying the Constraint

The primary capacity-constrained resource (CCR) is that resource which, more than any other, threatens the creation of throughput. If throughput is to be generated by the delivery of products to shipping, then it can be said that the primary CCR is that resource which threatens, more than any other resource, the ability of the system to deliver orders to shipping (see Fig. 1).

I. CHAPTER OBJECTIVES

- To present the Theory of Constraints (TOC) methodology for identifying the primary capacity constrained resource
- To introduce the role of the shipping buffer
- To create an understanding of the physical phenomena which impact the identification process

II. IDENTIFYING THE PRIMARY CCR

A. The Impact of Statistical Fluctuations on the Schedule

All resources are subject to statistical fluctuations in their ability to perform. In other words, sometimes an order will be early and sometimes it will be late. If an order becomes late due to a statistical fluctuation, to correct the problem additional capacity is required to bring the schedule back to its original plan.

The capacity required to correct problems in lateness due to normal statistical fluctuation is termed *protective capacity*. If there is not a sufficient amount of protective capacity to correct for lateness, each time a problem hits orders will become increasingly late.

Referring to Fig. 2, order A, originally planned for completion at 2:00, did not actually finish until approximately 2:25. A problem occurring on resource R-1 generated a delay of 25 minutes. This means that order B,

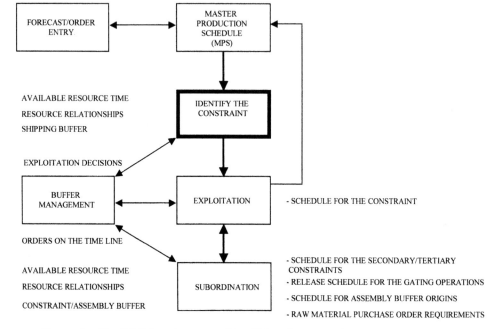

Figure 1 The TOC-based system—identifying the constraint.

originally planned to start at 2:00, was not able to start until 2:25. While B is processed in the proper amount of time, the 25 minute delay caused by the original problem has not been corrected. So order C is also started late. A problem again hits resource R-1 while processing order C. Order C, originally scheduled to be completed at 4:00, is not completed until 4:50 (a delay of 50 minutes).

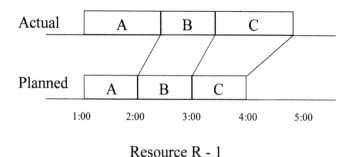

Resource R - 1

Figure 2 The impact of statistical fluctuation on the schedule.

Notice that there is a cumulative effect. Because resource R-1 does not have enough protective capacity to apply in correcting this problem each order is finished successively later.

This situation is exacerbated when more than one resource is involved in the process. Not only will orders processed on resource R-1 be chronically late, but this situation will cause orders on succeeding resources to be chronically late (see Fig. 3).

Resources R-1 through R-4 feed each other from left to right and are each loaded to 90%. In this case, the objective is to process inventory from resource R-1 to resource R-4 on time. But 90% is a relatively high load and the prospect of having a problem result in an order being late on resource R-1 is very high. The demand on resource R-2 is equally as great. Since resource R-1 will be chronically late, it is reasonable to consider that resource R-2 will not always be able to start on time. Since R-2's demand is high and yet it will always start late, the probability that resource R-2's orders will be even later is also high. This process will continue until reaching resource R-4.

The impact to the entire process can be readily visible if the load on each resource were to be converted to an 80% probability. In other words, there is an 80% probability of each resource completing its task on time. The combined probability for all resources in Fig. 3 is $.80^4$, or 41%. The string of resources must overcome a 59% probability that the order will be late.

An interesting side note is that the probability of lateness increases in proportion to exactly where in the string processing occurs. At 80%, the probability of lateness is greater for resource R-4 than it is for resource R-2.

It is this phenomenon which helps to identify the primary constraint. That resource which suffers from a lack of protective capacity more than any other has definitely been identified as the prime candidate.

B. The Role of the Shipping Buffer

As discussed in Chapter 5, a buffer is a time mechanism which acts to aggregate the time required to offset for those things which can go wrong. The shipping buffer offsets for those resources which exist between the primary constraint and shipping. Each order is protected so that no matter what may go wrong most orders will reach shipping on time.

90%   90%   90%   90%

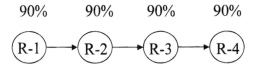

Figure 3   The lack of adequate protective capacity.

Those resources which, on the whole, suffer from a lack of adequate protective capacity will tend to have a cumulative effect on the time buffer. So one issue in identifying the constraint is to identify that resource which has the greatest impact on the time buffer in front of shipping.

Referring to Fig. 4, R-1 has five orders, A through E. Each order has been placed in time so that there is sufficient protection to ensure that they will usually make it to shipping on time. Notice that shipping is to the right. This is also the direction in which the time buffer used to protect shipping has been placed.

C. Applying Excess Demand

To determine the impact of demand on available protection means applying the time allocated to the orders against the time allocated to protection. However, there is still another issue which plays a part. Simply applying the load by dividing capacity by demand ignores the fact that the time to the right of each order is not available to be applied. To move the excess load to the right would immediately invalidate the delivery to shipping without first looking for more capacity. The only direction in which to apply the excess load is to the left, moving toward time zero (see Fig. 5).

After the demand has been loaded to a given day on the planning horizon and there is no more capacity available on the resource for that day, the remaining time is applied to the previous day.

In Fig. 6, after applying the excess demand to the left (Fig. 5) there is still space available before running into time zero (today). Since orders will not need to be displaced and each has its full protection (protection being the

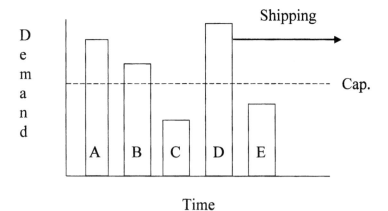

Figure 4 Protective capacity.

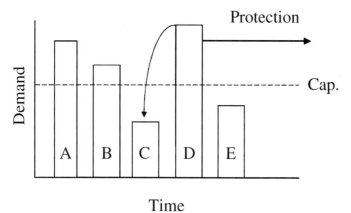

Figure 5 Protective capacity—excess load is applied to the left.

length of the shipping buffer), there is more than enough capacity available to reach shipping on time.

Resource R-2 has a similar situation. Notice that R-2's excess demand has been pushed to, but not past, time zero (see Fig. 7). R-2 will still have all its available protection and will not invalidate any shipping dates. The original protection that was applied to the right still exists.

However, any further increase in the load will begin to impact the amount of protection for those orders which must cross resource R-2.

D. Consuming Protective Capacity

Notice that R-3's orders have crossed time zero. This means that after applying the required protection to all the orders that must be shipped, there is not enough time available. Since the orders cannot be scheduled past time zero (a physical impossibility), protective time must be consumed to produce R-3's demand (see Fig. 8).

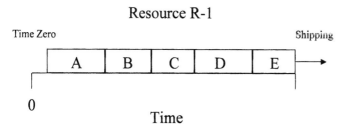

Figure 6 Protecting the shipping date.

Resource R-2

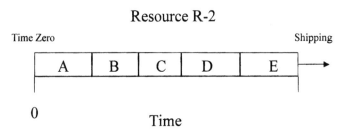

Figure 7 Protecting the shipping date.

At this time the primary constraint has yet to be identified. Notice that R-4 has crossed time zero to a greater extent than has R-3 (see Fig. 9). This means that it has even less protective capacity available than R-3. In fact, the demand on R-4 may be large enough that there is not even productive capacity available for some orders.

Among R-1, R-2, R-3, and R-4, the leading candidate to select as the primary constraint is R-4.

Note: This entire concept will be covered in detail later in this chapter.

III. CREATING THE DATA REQUIREMENTS

A. The Master Production Schedule

As discussed in Chapter 5, the identification process begins with input from the master production schedule (MPS). It is what drives the initial placement of orders on the timeline for each resource. Figure 10 represents the master production schedule, including demand origination, end item part numbers, quantity, and due date. Notice that product A's demand in period 100 originates from sales order S/O111. Notice also that demand for A in period 101 originates from the forecast.

Resource R-3

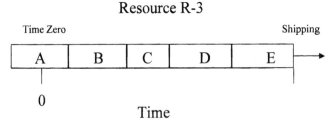

Figure 8 Consuming protective capacity.

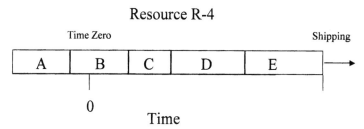

Figure 9 Identifying the primary constraint.

B. The Shipping Buffer

As discussed earlier, the *shipping buffer* is used during the identification phase to protect the shipping date, plan the initial placement of orders on the timeline during constraint identification, and determine the release time for material which does not go through the constraint. Orders are initially placed using the due date minus the shipping buffer. Figure 11 shows the three types of buffers needed to produce the schedule. Notice that the shipping buffer has been set at 8 hours. During the identification phase each order would be offset by 8 hours prior to being placed on the timeline. This is done in preparation for the creation of the schedule.

C. The Net

It is necessary to understand the following:

- Which part/operations will needed to be performed
- In what order the resources are to be used
- Amount of run time needed

DEMAND	PART	QUANTITY	DUE DATE
S/O111	A	20	100
S/O222	F	10	100
S/O333	F	20	101
FORECAST	A	20	101

Figure 10 The master production schedule.

BUFFER DATA

TYPE	LENGTH
SHIPPING	8 HOURS
CONSTRAINT	16 HOURS
ASSEMBLY	8 HOURS

Figure 11   Buffer data.

- Amount of inventory to apply so that downstream part/operations will not duplicate material that has already been produced

Figure 12 represents the net that is associated with the master schedule presented in Fig. 10. Since there are two items on the master schedule, the net used earlier has been expanded to include product F. *Note*: To save space within the net shown, setup time has been placed with the resource data.

D. Resource Data

Completing the assemblage of input to the identification phase will require resource capacity data and setup times (see Fig. 13).

Resource R-1 has 480 minutes per day of productive time and a quantity of one machine. Resource R-4 has 960 minutes per day available with a quantity of two machines.

IV. PLACING ORDERS ON THE TIMELINE

A. Allocating Inventory

Inventory is allocated first come, first serve. Those orders which must be completed first should have priority during processing. Allocating material begins by taking the earliest order in the planning horizon and processing it down the net to determine material requirements.

B. Processing on the Net

Since processing begins with the earliest order, from the master schedule 20 product As are due in period 100. According to the net in Fig. 12, the first part/operation to be processed is A/30 on resource R-1. Since there is no inventory

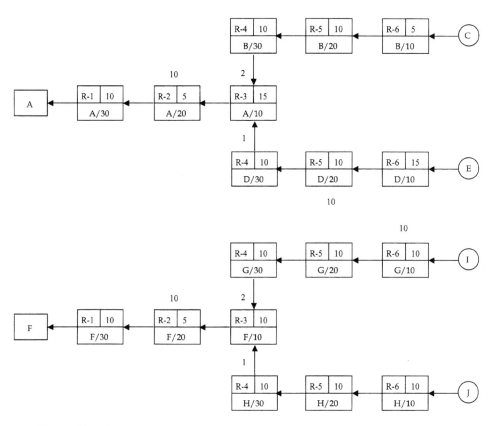

Figure 12 The net.

at A/30 to be used the entire 20 parts must be processed at 10 minutes each. Setup for resource R-1 includes another 10 minutes per order. After offsetting for the shipping buffer by 8 hours, this places a load of 210 minutes on resource R-1 on day 99 (see Fig. 14).

Part/operation A/20 is next in line to be processed. Notice that there are ten part/operation A/20s available in inventory to be applied toward the total demand of 20 units created by A/10's requirements. Each of the 10 remaining A/20s to be built requires 5 minutes of run time and 15 minutes of setup for a total time of 65 minutes (see Fig. 15). This time is again placed at the end of the shipping buffer. Notice that in going from A/10 to B/30 within the net there is a bill of material quantity of two B/30s needed for every A/10 produced. Bill of material relationships are taken into consideration when processing down the net. B/30's requirements include

RESOURCE DATA

RESOURCE	CAPACITY	S/U	QUANTITY
R-1	480	10	1
R-2	960	15	2
R-3	480	10	1
R-4	960	10	2
R-5	480	15	1
R-6	480	10	1

Figure 13   Resource data.

run time for a quantity of ten times the bill of material quantity of two. Twenty pieces are processed at 10 minutes each plus a setup of 10 minutes for a total of 210 minutes (see Fig. 16).

Notice that the timeframe for each part/operation remains the same for each load placed on each resource. At face value this does not seem logical. How can the load for successive operations appear in the exact same place in time. Remember that *the object of this exercise is not to schedule the factory but to determine which resource is the biggest threat to the creation of throughput.* Remember also that the optimal transfer batch size is quantity one. Since the processing of one part in comparison to the total protective time used in the factory is so small, the difference between

Resource R-1

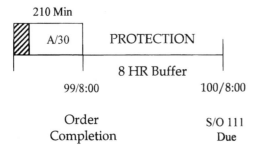

Figure 14   Processing on the net—resource R-1.

Resource R-2

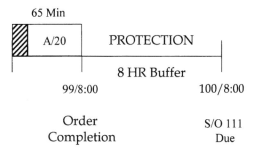

Figure 15 Processing on the net—resource R-2.

where a part would fit on the timeline by including the actual processing time or sequencing is minuscule.

Each part/operation within the net is processed similarly for all orders on the MPS. It stops when reaching the last part/operation. While the purchased part quantity requirements are computed during this time, no raw material order timing will be generated until the subordination phase.

C. Computing Start and Stop Times

Figure 17 is a compilation of the data used in computing start and stop times for the initial placement of all orders on the timeline for resource R-4. Notice that sales order S/O222 for a quantity of ten product Fs generated zero time. This is due to the inventory available at part/operation F/20.

Resource R-4

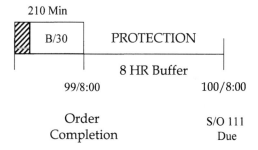

Figure 16 Processing on the net—resource R-4.

Sales Order/Qty	Order Due	Part/ Oper.	Proc. Qty.	Setup Time	Run Time	Total Time	Start Time	End Time
S/O111/20	100	B/30	20	10	10	210	99/4:30	99/8:00
		D/30	10	10	10	110	99/6:10	99/8:00
S/O222/10	100	G/30		10	10	0		
		H/30		10	10	0		
S/O333/20	101	G/30	40	10	10	410	100/1:10	100/8:00
		H/30	20	10	10	210	100/4:30	100/8:00
Forecast/20	101	B/30	40	10	10	410	100/1:10	100/8:00
		D/30	20	10	10	210	100/4:30	100/8:00

Figure 17 Computing start and stop times.

The initial placement of B/30 for sales order S/O111 is at the end of day 100 minus the shipping buffer, or hour 8:00 on day 99. The start time is set at 4:30 on day 99, a total of 210 minutes earlier than the stop time. The 210 minutes is computed by multiplying the quantity of 20 part requirements by the run time of 10 minutes and adding the 10-minute setup. Once the stop time of the order has been computed, to find the start time simply subtract the total processing time from the stop time.

D. Accumulating the Demand

As the process continues down the net for each successive order on the master schedule, demand is accumulated for each resource (see Fig. 18). Order S/O111 generates demand for B/30 of 210 minutes and demand for D/30 of 110 minutes to be completed on resource R-4 8 hours prior to the end of ship date 100. So the load for each is placed at the end of day 99 at hour 8:00.

Sales order S/O222 will be filled by inventory which exists at part/operation F/20. Since this part/operation exists prior to part/operations G/30 and H/30, no demand will be generated.

The next order on the master schedule is S/O333. This order generates 410 minutes on resource R-4 to create part/operation G/30 and 210 minutes to create part/operation H/30.

The last order generated is to fill a forecast for 20 As on day 101. This places 410 minutes demand on part/operation B/30 and 210 minutes on D/30

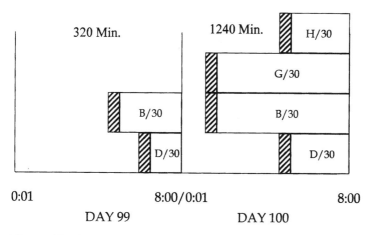

Figure 18   Accumulating demand.

at the end of day 100. The total load on resource R-4 is 1240 minutes for day 100 and 320 minutes for day 99.

E.   Pushing the Load Toward Time Zero

The next step in the process, after all resource labor is placed on the timeline for every part/operation, is to level the load. Since the protection exists to the right, the process of leveling the load begins at the end of the planning horizon and moves constantly toward time zero (to the left). The original protection generated at the time the orders were placed on the timeline will thus still be available.

Notice that resource R-4 has two machines. This means that two orders, one on each machine, can be run concurrently.

The first part/operations to be placed are B/30 and D/30 from the forecast (see Fig. 19). Part/operations B/30 and D/30 were placed so that their completion times were the same as those for the original placement on the timeline. There were no additional machines on which to place H/30 and G/30 for sales order S/O333, so they needed to be moved. Part/operations B/30 and D/30 from sales order S/O111 were repositioned last.

Notice that B/30 and D/30 for sales order S/O111 did not include setup time. Since setup savings can have a very large impact on finding additional capacity, for the purposes of identifying the constraint addi-

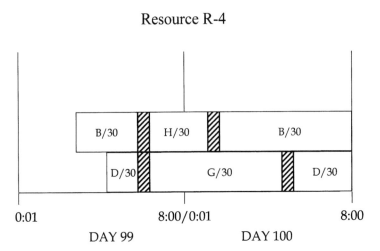

Figure 19 Pushing the load toward time zero.

tional setups are left out. One setup for part/operations B/30 and D/30 had already been placed.

F. Recomputing Start and Stop Times

In recomputing start and stop times for G/30 and H/30, they were placed immediately after D/30 and B/30, respectively (see Fig. 20).

Part/operation D/30 from the forecast has a start time of hour 4:30 on day 100, so G/30 must stop at 4:31 on day 100. After subtracting the 410 minutes for processing time, G/30's start time is placed at 5:41 on day 99.

Part/operation B/30's start time is hour 1:10 also on day 100, so H/30 must stop at 1:09 day 100. After subtracting 210 minutes, H/30's start time is computed as 5:39 on day 99.

Part/operation D/30 for S/O111 is placed after G/30's start time of 5:41 day 99. Its stop time is therefore 5:40 on day 99. After subtracting 100 minutes for processing time, D/30's start time is placed at 4:00. (Notice that the time for the second placement of a D/30 part/operation did not include setup time.)

The last part/operation to be repositioned is B/30. Its stop time is placed immediately after H/30's stop time of 5:39 day 99. Its start time is placed 200 minutes earlier at 2:18 day 99.

G. Consuming Protective Capacity

To demonstrate the consumption of protective capacity, machine 2 will be removed from resource R-4, cutting its capacity in half. The load requirements will remain the same (see Fig. 21).

Sales Order/Qty	Order Due	Part/ Oper.	Proc. Qty.	Setup Time	Run Time	Total Time	Start Time	End Time
S/O111/20	100	B/30	20	10	10	210	99/2:18	99/5:38
		D/30	10	10	10	110	99/4:00	99/5:40
S/O222/10	100	G/30		10	10	0		
		H/30		10	10	0		
S/O333/20	101	G/30	40	10	10	410	99/5:41	100/4:31
		H/30	20	10	10	210	99/5:39	100/1:09
Forecast/20	101	B/30	40	10	10	410	100/1:10	100/8:00
		D/30	20	10	10	210	100/4:30	100/8:00

Figure 20 Recomputing start and stop times.

Time zero has been placed at just over halfway through day 98. This happens to coincide with the beginning of the run time portion of part/ operation H/30, which means that all the setup time for H/30 and the run time for D/30 as well as B/30 must be run in the past. Obviously, this cannot occur and the orders must be adjusted. But in comparing one resource with another, there must be a number which represents the extent to which each resource has passed time zero. The time allocated to the orders which have passed time zero must be accumulated. In the case of resource R-4, the load has passed time zero by the setup time for H/30 (10 minutes) plus the run time for D/30 (100

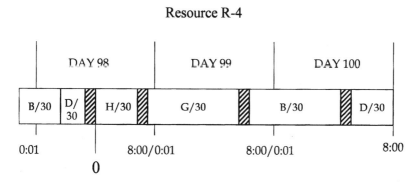

Figure 21 Consuming protective capacity.

Sales Order/Qty	Order Due	Part/ Oper.	Proc. Qty.	Setup Time	Run Time	Total Time	Start Time	End Time
S/O111/20	100	B/30	20	10	10	210	97/5:57	98/1:27
		D/30	10	10	10	110	98/1:28	98/3:18
S/O222/10	100	G/30		10	10	0		
		H/30		10	10	0		
S/O333/20	101	G/30	40	10	10	410	98/6:50	99/5:40
		H/30	20	10	10	210	98/3:19	98/6:49
Forecast/20	101	B/30	40	10	10	410	99/5:41	100/4:29
		D/30	20	10	10	210	100/4:30	100/8:00

Figure 22   Recomputing start and stop times.

minutes) plus the run time for B/30 (200 minutes) for a total of 310 minutes. This means that R-4's demand has taken 310 minutes of protective capacity away from the system.

   Note: While 310 minutes of protective capacity has been taken, the impact to individual orders is still uncertain. This can only be determined after pushing the load so that B/30's start time is equal to time zero.

H.   Recomputing Start and Stop Times for a One-Machine Scenario

Figure 22 is a representation of the new placement of each part/operation for the load on resource R-4. Notice that the start time of B/30 for sales order S/O111 has been set at 5:57 on day 97. Time zero is at hour 3:09 on day 98 and is actually set to the computer's current date and time at the beginning of the process.

   Unless there is another resource which pushes demand even farther past time 5:57 on day 97, resource R-4 would be considered the first candidate for the primary constraint.

   In order to ensure that resource R-4 performed to the maximum capable in generating throughput, it must be exploited. This would include generating a schedule to ensure that, if followed, orders would arrive to shipping on time.

Sales Order/Qty	Order Due	Part/ Oper.	Proc. Qty.	Setup Time	Run Time	Total Time	Start Time	End Time
S/O111/20	150	B/30	10	10	15			
		D/30	40	15	10			
S/O333/20	151	G/30	15	20	15			
		H/30	20	10	20			
Forecast/20	151	B/30	20	10	15			
		D/30	40	15	10			

Figure 23 Recomputing start and stop times. Shipping buffer is 12 hours.

V. CONCLUSION

When trying to identify those resources which have limited capacity to perform to the demands of the market, simply dividing capacity by demand will not accomplish the desired task. Given the physical issues which must be addressed a different approach should be used which assesses the amount of protection required and then applies the demand created by a given master schedule against the protection.

VI. STUDY QUESTIONS

1. What impact do statistical fluctuations have on the generation of a schedule?
2. What role does protective capacity play in the generation of a schedule?
3. What purpose does the shipping buffer serve?
4. When applying demand to the available capacity for a specific schedule generated by the master schedule, what is the traditional method used and why does it fail to accomplish its task?
5. Define what is meant by the term *time zero* and describe its function.
6. Describe the method used to identify the constraint using the TOC-based system.
7. What data elements are needed by the TOC-based system to identify the constraint?
8. What data elements are contained in the net?

9. What is meant by the term *consuming protective capacity?*
10. How does excess capacity manifest itself?
11. What is the purpose of productive capacity?
12. What should the goal of an information system be and why?
13. What is meant by the term *processing on the net?*
14. Using the data in Fig. 23, compute the start and stop times for the initial placement of orders on the timeline.
15. Once the start and stop times have been computed for Question 14, level the load, recomputing the start and stop times based on an 8-hour day. Assume only one machine is available. Time zero will be day 147.

7
Exploitation

The purpose of exploitation is to maximize the amount of throughput generated by the constraint. While this can mean a number of different activities, such as ensuring that the correct product mix is being sold into the market, for the purposes of this chapter, the exploitation process of scheduling will be the primary subject (Fig. 1).

I. CHAPTER OBJECTIVES

- To address the process of scheduling the constraint
- To demonstrate how the new system identifies possible trouble areas
- To show how to maximize the constraint's utilization

II. EXPLOITATION IN SCHEDULING

In scheduling the constraint, the objective is to maximize its time by ensuring that

- It is constantly operating to generate throughput.
- Whenever and wherever possible, additional constraint time is found.
- Those orders which may have particular problems are identified so that action can be taken.

A. Readjusting Orders on the Timeline

In Chapter 6, resource R-4 had been identified as a candidate for the primary constraint. Since the orders have already been placed on the timeline but are

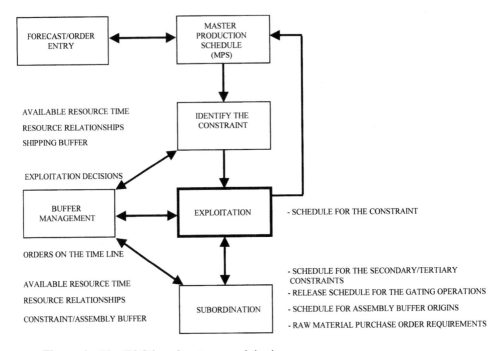

Figure 1 The TOC-based system—exploitation.

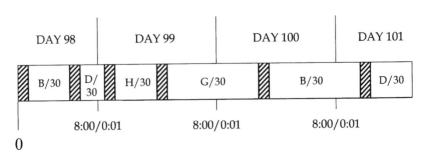

Figure 2 Adjusting orders on the timeline.

attempting to use time which no longer exists, the first step in the exploitation process would be to push the orders so that they are no longer in the past (see Fig. 2).

By moving the orders so that none appears prior to time zero, all orders on the timeline have been shifted to the right. Each order will now have a different start and stop time, which will be in this case 330 minutes later. The 330 minutes was determined by adding the setup time for H/30 (10 minutes) plus the setup and run time for B/30 (210 minutes) plus the setup and run time for D/30 (110 minutes). Notice that the setup times for D/30 and B/30 have been added back to the schedule.

Note: while 310 minutes of protective capacity has been taken from resource R-4, the impact to individual orders is still uncertain. This can only be determined after pushing the load so that B/30's start time is equal to time zero. A comparison can then be made between the stop date and time of the individual order and the due date and time of the sales order to the length of shipping buffer.

1. Recomputing Start and Stop Times

As mentioned, the new start date and times can be determined by adding 330 minutes to the start and end dates and times for each part/operation of resource R-4 (see Fig. 3).

Sales Order/Qty	Order Due	Part/ Oper.	Proc. Qty.	Setup Time	Run Time	Total Time	Start Time	End Time
S/O111/20	100	B/30	20	10	10	210	98/3:27	98/6:57
		D/30	10	10	10	110	98/6:58	99/0:48
S/O222/10	100	G/30		10	10	0		
		H/30		10	10	0		
S/O333/20	101	G/30	40	10	10	410	99/6:20	100/3:50
		H/30	20	10	10	210	99/0:49	99/6:19
Forecast/20	101	B/30	40	10	10	410	100:/3:51	101/2:01
		D/30	20	10	10	210	101/2:02	101/5:32

Figure 3 Recomputing start and stop times.

2.   Analyzing the Impact to Each Order

In recomputing the start and stop times for each order, each must be compared to the original placement to determine the true impact to the protection allocated to each order.

In Fig. 4, notice that B/30's total buffer is 14 hours and 57 minutes and is over the requirement of 8 hours by 6 hours and 57 minutes. D/30 also has an excess in buffer time of 7 hours 17 minutes. This means that S/O111 will have no problems in meeting its due date requirements.

Order S/O333 should also have no problems in meeting its due date. However, the forecast of 20 part As is in jeopardy. While part/operation B/30 still has over 50% of its buffer remaining and should be on time, it must be matched with part/operation D/30. Part/operation D/30 has only 2 hours and 28 minutes, or 34% of its original protection, available. The probability of D/30 being completed on time is extremely low due to the lack of adequate protection. The forecast of 20 product As will be late.

3.   Adjusting for Inventory

Notice that only one of the part/operations (D/30) scheduled to cross R-4 has inventory available to begin processing on the portion of the net for product A (Fig. 5). But B/30 is scheduled first. This means that if resource R-4 were to be the primary constraint, throughput would drop immediately unless the schedule were changed to process part/operation D/30 first.

Sales Order/Qty	Order Due	Part/ Oper.	Buffer Due	End Time	Buffer Avail	Buffer Req.l	Delta
S/O111/20	100	B/30	99/8:00	98/6:57	14:57	8 hours	+6:57
		D/30	99/8:00	99/0:48	15:17	8 hours	+7:17
S/O222/10	100	G/30					
		H/30					
S/O333/20	101	G/30	100/8:00	100/3:50	12:19	8 hours	+4:19
		H/30	100/8:00	99/6:19	14:19	8 hours	+6:19
Forecast/20	101	B/30	100/8:00	101/2:01	5:59	8 hours	-2:01
		D/30	100/8:00	101/5:32	2:28	8 hours	-5:32

Figure 4   Analyzing the impact to the protection allocated to each order.

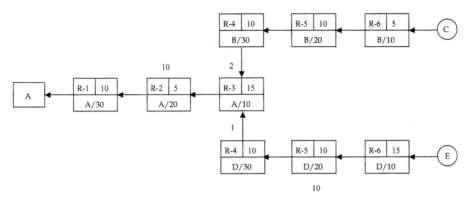

Figure 5 Adjusting for inventory—the net.

Since no other part/operations for any other resource have been scheduled, there will not be a product mix problem. Other resource requirements must be determined by the demand created by the constraint, and unless a finalized schedule has been created, product mix is not an issue at this time (see Fig. 6).

Notice that D/30 for sales order S/O111 is now scheduled to be run on resource R-4 first. The resulting changes to the start and stop times are affected for D/30 and B/30 only (see Fig. 7). The previous example isolated one order with material to move to the front of the schedule. To ensure that the constraint is protected in reality, enough orders with material should be in front of the constraint to cover two-thirds the length of the constraint buffer.

Resource R-4

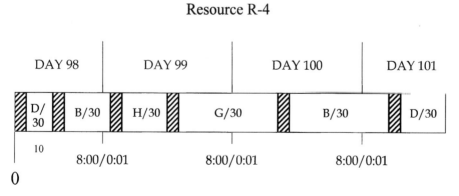

Figure 6 Adjusting for inventory—resource R-4.

Sales Order/Qty	Order Due	Part/ Oper.	Proc. Qty.	Setup Time	Run Time	Total Time	Start Time	End Time
S/O111/20	100	D/30	10	10	10	110	98/3:27	98/5:17
		B/30	20	10	10	210	98/5:18	99/0:48

Figure 7 Adjusting start and stop times.

B. Exploiting the Constraint Through Setup Savings

There are a number of ways in which D/30's problem can be fixed. One is the possibility that by combining the orders for D/30 and B/30 from the forecast with those from sales order S/O111 additional time can be found by saving setup. The key issue here is to ensure that no other problems will be created in the process.

Since there is only 20 minutes of setup time that can be gained and the amount of buffer that has been taken by the schedule for part/operation D/30 is 5 hours and 32 minutes, it seems unlikely that a solution can be found. However, Fig. 8 shows the impact of setup savings.

Notice that B/30 and D/30 from the forecast have been combined with D/30 and B/30 from S/O111 to save setup. Notice also that G/30 and H/30 have been pushed out to later dates by the amount of displacement caused by the run times for the forecasted D/30 and B/30 part/operations (see Fig. 9).

The completion time for G/30 is now set at 101/5:12 and 20 minutes of constraint time has been saved, which can be used to create other products. However, the lateness of order S/O333 may be a problem for the customer.

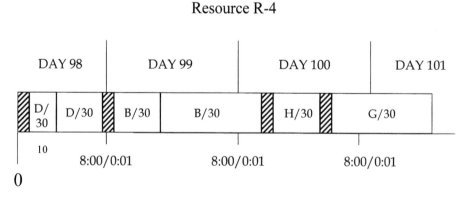

Resource R-4

Figure 8 The impact of setup savings.

Sales Order/Qty	Order Due	Part/ Oper.	Proc. Qty.	Setup Time	Run Time	Total Time	Start Time	End Time
S/O111/20	100	D/30	10	10	10	110	98/3:27	98/5:17
Forecast/20	101	D/30	20	10	10	200	98/5:18	99/0:38
S/O111/20	100	B/30	20	10	10	210	99/0:39	99/4:09
Forecast/20	101	B/30	40	10	10	400	99/4:10	100/2:50
S/O222/10	100	G/30		10	10	0		
		H/30		10	10	0		
S/O333/20	101	G/30	40	10	10	410	100/6:22	101/5:12
		H/30	20	10	10	210	100/2:51	100/6:21

Figure 9 Computing start and stop times for setup savings.

Unless the customer can live with the order being late it may be best not to use setup savings. In addition, the forecast has yet to be sold. Processing it instead of products that can generate immediate throughput is not the wisest choice.

Setup savings can involve a number of issues. If the size of individual setups is large or the number of setups within the planning horizon is high, then setup savings can be a great way of gaining additional constraint time. However, in setup savings it is important to remember that whenever two orders are combined, another order's start date is postponed.

1. The Impact of Setup Savings on Other Resources

Setup savings can have a positive or negative impact on any given resource. Once setup savings has been performed, those resources which exist prior to the scheduled resource must perform to the new demand created by setup savings. This may not be good for those resources already approaching constraint levels.

Figure 10 represents a chain of resources starting with resource 1 on the left. Each resource feeds the next in succession until finally reaching

Figure 10 The impact of setup savings.

resource number 4. The capacity of resource 4 is 50 units per hour and is loaded to 100%.

When setup savings is performed as in Fig. 8, more units can be produced in less time. Some of the capacity that was dedicated to performing setup activity will now be dedicated to productive activity.

Notice that resource 3 is loaded to 90% when resource 4 is producing at 50 units per hour. What will be the impact on resource 3 if, through setup savings, resource 4 now produces 60 units per hour? The demand on resource 3 will increase. Since it is already loaded to 90% of its rated capacity, resource 3 will probably become a secondary constraint. As a result, a schedule must be built for resource 3 maximizing its utilization so that 4's new production rate can be protected. Along with the creation of the schedule a buffer must also be created. An additional buffer means an increase in inventory in front of the secondary constraint.

The situation as described is not necessarily bad. Remember that because resource 4 is now producing more, additional throughput is being generated to offset the additional inventory. The real problem is created if resource 3 cannot be subordinated to the schedule created on resource 4.

2.   Limiting the Extent of Setup Savings

If for every order that is pulled in and combined other orders are displaced, then it may be wise to limit the number of orders which can be combined. As an example, in Fig. 11, the top grouping of orders represents the schedule prior to setup savings. In the second grouping three As are combined, one from the beginning, one from the middle, and one from the end of the original schedule. The amount of time saved was approximately equal to the amount of setup and run time for one A. However,

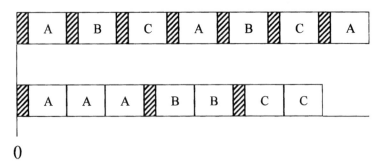

Figure 11   Setup savings.

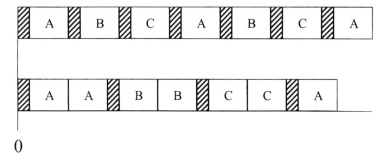

0

Figure 12 Limiting the amount of setup savings.

notice how far to the right the first C has been moved. Its position is now close to the end of the schedule.

The questions that must be asked during setup savings are

- How far into the schedule should the process go?
- Should there be a limitation?
- If so, what should the limitation be?

One method is to use a number that can relate the amount of setup time saved with the amount of time into the future to look for additional orders. As an example, the number 25 would mean that for every hour saved the system might look into the future by 25 hours. As seen in Fig. 11 it is important to limit the amount of setup savings that can be used in the planning horizon to some adjustable number (see also Fig. 12).

Notice that the last A in the schedule was not pulled in to be combined with the order for the first and second A. Instead, only two As were combined. Notice also that the first order of C was not pushed out as far.

Note: One rule to remember in executing setup savings is not to make matters worse than they already are.

C. Looping Back to the MPS

Since a decision has been made not to perform setup savings and the forecast for product A is going to be late, when should the MPS be told to expect this product to be available to promise (ATP) (if forecasted) or shipped (if a sales order)?

Once an actual due date for the part/operation has been established for the constraint, it is the due date plus the shipping buffer which establishes the new date (see Fig. 13). The master schedule would be instructed to reschedule the forecast for 20 As from 101/5:32 (the due date of the

Resource R-4

Forecast/20	101	B/30	40	10	10	410	100:/3:11	101/2:01
		D/30	20	10	10	210	101/2:02	101/5:32

Part Operation Due + Shipping buffer = New Delivery Date/Time

$$101/5{:}32 + 8 = \mathbf{102/5{:}32}$$

Figure 13   Looping back to the MPS.

order across the constraint) to 102/5:32 (the constraint's due date plus the 8-hour buffer).

D.   Identifying Late Orders

Sales orders are considered to be having serious timing problems whenever the due date of the order crossing the constraint uses over 50% of the buffer in front of shipping. In other words, if the due date of the order is greater that the ship date of the sales order minus one-half the shipping buffer, it will be late.

   Whenever the buffer is being used at less than 50% but more than zero it is considered a potential problem, but no action is usually taken.

1.   Late Orders List

The late orders list is a feature of the system and shows those orders which are going to be late according to their original due dates and how many days they will be late (see Fig. 14).

Part/Op.	QTY	Sales Orders	Sales QTY	Due	Days Late
A/30	100	S/O111	50	152	2
		S/O222	50	153	1
B/30	150	S/O333	150	154	5
B/30	50	S/O111	50	154	5

Figure 14   The late orders list.

E. Applying Overtime

The way overtime is applied can have three effects:

1. If begun too early in the process, it will drive inventory up.
2. If begun too late in the process, it will increase operating expense without maximizing the throughput generated.
3. If done at the right time and place, inventory and operating expense can be held to a minimum while maximizing the amount of throughput generated.

Any time a part/operation will be late overtime is a valid option, but only after all attempts have been made to gain more capacity (as in setup savings) without spending more money. In attempting to maximize throughput at the constraint prior to spending overtime, the constraint may be sufficiently exploited and overtime may not be needed. Whenever authorized, overtime should be used sparingly and only at the right time and place.

It's not enough to know that the schedule is late. Specifically if it can be fixed, when and where to place overtime and what orders are being threatened are major concerns. If it is possible to pinpoint exactly when overtime should be worked, what resources are involved, and what part/operations are to receive the overtime effort, then applying the additional expense will have a direct effect on increasing throughput while keeping inventory and operating expense to only that required.

In Fig. 15, the two shaded orders (3 and 6) are indicating that they are using over 50% of the buffer and that they will be late unless something is done. All attempts so far to gain additional capacity have failed to solve this problem, so overtime is suggested. The question left unanswered is where must the overtime be applied to maximize effectiveness.

The closer to time zero that overtime is applied, the greater the number of orders it will affect. As an example, if overtime were applied to order 1, it would impact the due date of every order in the schedule by completing them earlier. It would also require that more inventory be made available for processing.

Resource R-1

Figure 15 Overtime assignment.

If overtime were applied to order 7, it would impact only order 7. The amount of additional throughput generated would therefore be minimized. If overtime were applied to order 2, inventory would be held to the minimum required, while throughput would be maximized in that all orders to the right of order 2 would be impacted favorably. It is only after expending all allowable overtime on order 2 and still showing late orders that order 1 be allowed to absorb additional overtime.

If the scheduling problem still persists, the next 2-hour block of overtime would be applied to order 5, followed by an equal amount applied to order 4. This process of jumping forward to add overtime to the order just in front of the late order continues until one of the following occurs:

- All late orders have been fixed.
- All available overtime has been expended.
- The end of the planning horizon is reached.

There is an important exception to this process. Overtime is usually spent at the end of the day. If order 2 existed in the same day as order 3, then the overtime would come too late. It would be expended at the end of the day, and since order 3 comes after order 2 the overtime would be placed after order 3 begins and ends. This means that no change will be made in order 3's timing.

So there must be a modification to the rule. Overtime must be placed at the end of the day prior to the late order and should be expended first on those orders which are closest in time, but exist prior, to the late order.

1. Limiting the Amount of Overtime

Overtime is usually limited in the system based on the amount of overtime to be applied per day or week. If overtime were limited to 2 hours but this was not enough to make orders 3 and 6 finish on time, then more overtime must be spent. Two hours was originally spent on order 2. Since this did not solve the problem, an additional 2 hours could be applied to order 1.

Sales Order/Qty	Order Due	Part/ Oper.	Proc. Qty.	Setup Time	Run Time	Total Time	Start Time	End Time
Forecast/20	101	B/30	40	10	10	410	100:/3:11	101/2:01
		D/30	20	10	10	210	101/2:02	101/5:32

Figure 16   Recomputing start and stop times.

Sales Order/Qty	Order Due	Part/ Oper.	Proc. Qty.	Setup Time	Run Time	Total Time	Start Time	End Time
Forecast/20	101	B/30	40	10	10	410	100:/3:11	101/0:01
		D/30	20	10	10	210	101/0:02	101/3:32

Figure 17 Placing the overtime.

2. Recomputing Start and Stop Times

The original schedule for D/30 in the forecast placed it 5 hours and 32 minutes into the 8-hour buffer (see Fig. 16). In order to solve this problem, 2 hours of overtime is applied at the end of day 100 to part/operation B/30 (see Fig. 17).

Notice that part/operation B/30 now finishes 2 hours earlier, allowing D/30 to start and finish earlier. Since D/30 now finishes at 101/3:32, it uses less of the buffer (44%) and should reach shipping on time. No additional overtime is required. Notice, however, that this order is marginal. If a large percentage of orders will use buffer time while generating the schedule, the cumulative effect can have a negative impact to on-time delivery.

3. Manual Overtime Assignment

While the system should provide the ability to automatically add overtime to the appropriate resource, it will be advantageous to also provide the ability to

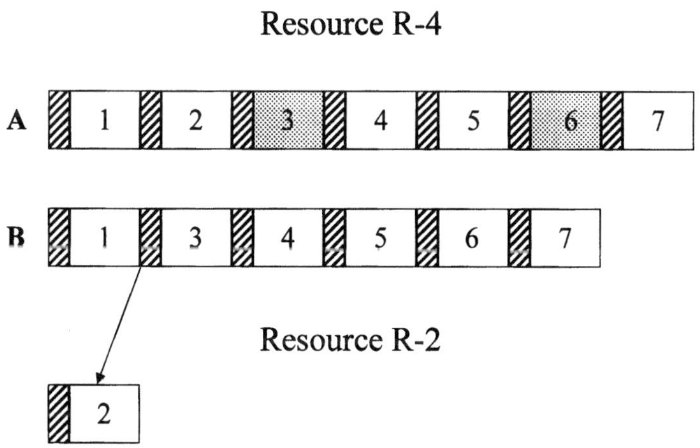

Resource R-4

Resource R-2

Figure 18 Off-loading to alternative resources.

Figure 19   The impact of off-loading on the subordination process.

apply additional overtime manually. When reviewing orders on the timeline the system should provide the ability to select specific orders on which to spend the additional overtime.

F.   Off-Loading

If the attempt to gain more capacity through setup savings fails, it might be a good idea to off-load an order to another resource which is capable of handling it (see Fig. 18). Schedule A for resource R-4 indicates that orders 3 and 6 are late. Notice that order number 2 is no longer on schedule B and that orders 3 and 6 are now on time as indicated by the shading having been removed. Order 2 was chosen because it was early in the process and, as discussed, will impact a majority of orders. When choosing an order to off-load keep in mind that the object of the exercise is to improve the situation, not cause more problems. Choose an order, or a number of orders, which exists early in the schedule, approximately equals the total amount of late buffer minutes, and for which there is another resource that will not be negatively affected.

While it is not always possible to find additional capacity by off-loading to a different resource, it can be a very effective alternative. However, the only way to understand the impact of an order which has been off-loaded is through attempting to subordinate the resource on which the order is now being processed with the schedule currently being created for the primary constraint. It may be found that it is impossible to schedule the remaining resources to the additional capacity generated by the off-load (see Fig. 19).

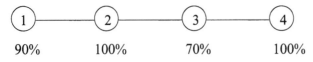

Figure 20   The impact of off-loading on the subordination process—situation after orders from resource R-4 are off-loaded to resource R-2.

Resource 4 is loaded to 120% and is the primary constraint. Scheduling and setup savings have not been able to reduce the load adequately. A decision has been made to off-load orders from resource 4 to another resource. The only resource capable of handling the requirements demanded to create the part is resource 2, which is loaded to 80%. When the load is taken from resource 4 and placed on resource 2 the demand for resource 2 increases to 100% (see Fig. 20).

This means that the string of resources needed to produce the product has three resources in the chain of four which are over 90%, and two are loaded to 100%. The chance of being successful with resources scheduled to this extent seems remote. However, to ensure that off-loading from resource 4 to resource 2 is viable will require creating a secondary schedule for resource 2. If any conflicts exist between the schedules, then off-loading to resource 2 will probably not be a good alternative.

1. Lot Splitting

During the off-loading process it may be desirable to off-load less than an entire batch. It may be that in off-loading less than the entire amount the user is able to prevent problems occurring while attempting to subordinate to the constraint's schedule (see Fig. 21).

Notice that order 2 now appears on both resource R-4 and R-2 and that setup times exist for both lots. While the total time for processing order 2 will increase, if resource R-2 has enough capacity, then throughput will go up and there will be no impact to inventory or operating expense.

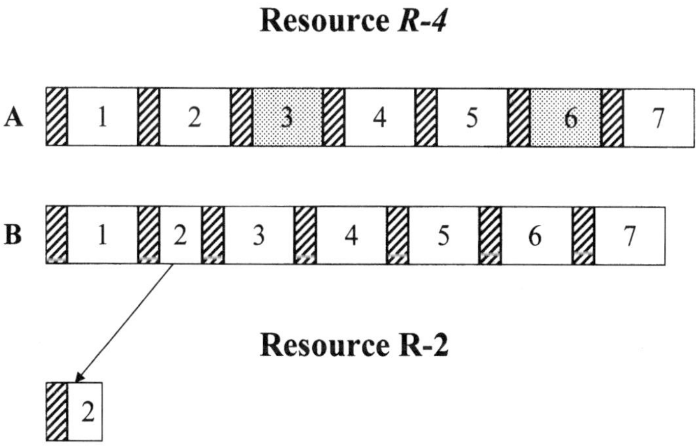

Resource *R-4*

Figure 21 Lot splitting.

III. PROBLEMS WHEN THE CONSTRAINT FEEDS ITSELF

Often there exists a problem where an order which must be processed on the constraint at one part/operation must be processed again at a later part/operation. This order actually appears on the schedule for the constraint in two places, once for the earlier part/operation and again for the second.

If an order is to appear in two places on the same schedule, there must be time allotted for the order to be processed at the first part/operation, continue to other resources, and then return. But remember that when each order was placed on the constraint an estimate of the time buffer needed was applied equally to all orders. Unless something is done to change this, both part/operations will be scheduled to cross the constraint, at least initially, at the same time. When the load is leveled this will change, but there is no guarantee that the part/operations will not appear next to each other on the constraint schedule. This is a problem. There must be a mechanism to ensure that when an order has two part/operations which cross the constraint that they appear in the correct sequence and that they are kept a minimum distance from each other (see Fig. 22).

In schedule A, part/operations 3 and 4 belong to the same sales order. Each must be processed on the constraint, but after 3 is processed it must leave the constraint to receive processing at other resources and then return as part/operation 4. Since the stop time for 3 is almost identical with the start time for 4, the likelihood of having the necessary procedures completed are nil.

What is needed is some kind of a buffering mechanism that will ensure that these part/operations do not appear too close together within the

Resource R-4

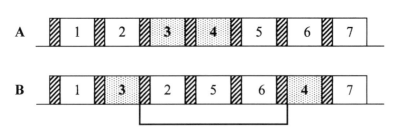

Figure 22   The rods concept—orders 3 and 4 must be held apart to keep them from being adjacent to each other on the constraints schedule.

schedule for the constraint (see schedule B). Notice that 3 and 4 have a certain amount of protection, which has been applied to them so that they do not appear together on the constraint's schedule.

A. Rods

Rod is the term given to the buffering mechanism briefly discussed in the last section. There are two situations in which rods become necessary:

- Whenever there is a connection between part/operations on the same resource schedule, as seen in Fig. 22
- Whenever a connection exists between the schedule of part/operations on multiple resources

There are also a number of different types:

- Forward rods
- Backward rods
- Combination rods

Forward rods are used to keep part/operations from advancing too far forward and extend from the front of one part/operation on the schedule to the back of the next one. Backward rods extend from the back of a part/operation to the front of a preceding part/operation. Combination rods do both.

There can be opportunities for many conflicts requiring this kind of resolution within the same or multiple schedules. However, what will be discussed here will be those conflicts and the rods which apply to a single schedule. Multiple schedule conflicts will be discussed in Chapter 8, "Subordination."

Resource R-4

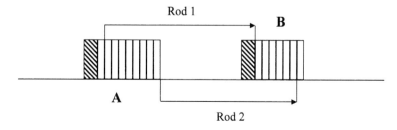

Figure 23 Forward rods.

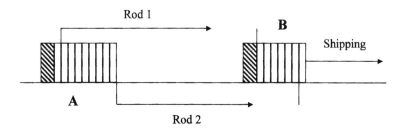

Figure 24   Forward rods—B may be shifted to right only.

1.   Forward Rods

Figure 23 represents two forward rods. Notice that there are two rods extending from part/operation A to part/operation B. The objective is to ensure that when these two part/operations are placed on the timeline and during the subsequent scheduling process that the first part finished at part/operation A will be completed by a minimum amount of time before the start of the first part at part/operation B. Additionally, the last part at A should finish by a minimum amount of time before the last part at B begins.

It is not necessary that A finish exactly at a certain distance from B. It may finish earlier (see Fig. 24), but it should never finish less than the minimal allowed distance. In the order of initial placement, B would be placed on the timeline first based on the shipping buffer and, since A must finish before B, it should be placed a minimum of one-half the constraint buffer in the direction of time zero. When the orders are leveled during the

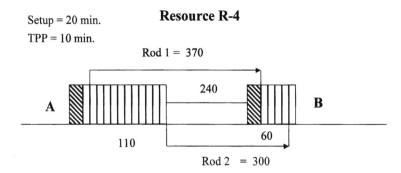

Figure 25   Computing the length of the forward rods.

identification process, A might be pushed further away from B. It should never be moved closer.

2. Computing the Length of the Rods

Obviously, the rods are a simple method of describing the buffering activity graphically and do not exist inside the computer in anything more than mathematical computations (see Fig. 25).

The length of one-half the constraint buffer is 240 minutes. Setup time is 20 minutes for both part/operations and run time is 10 minutes per piece. Each of the small nonstriped blocks represents a part. To compute rod 1 the amount of run time for all but the first part of part/operation A (110 minutes) is added to the length of the buffer (240 minutes) plus the setup for part/operation B (20 minutes). The length of rod 1 is 370 minutes.

Rod 2 is computed as the length of the buffer (240 minutes) added to the setup time for part/operation B (20 minutes) plus the run time for all but the last part. The length of rod 2 is 300 minutes.

3. Backward Rods

Backward rods move from the latest part/operation to the earliest. Notice in Fig. 26 that rod 1 originates at the end of setup for B and stops at the stop time for the first part for A. Rod 2 starts at the beginning of the last part B and stops at the end of the last part A. The objective here is that if B must move to the left, it will push part/operation A into an earlier time (to the left).

The lengths of the backward rods are computed similar to the forward rods. Rod 1 equals the setup time for B (20 minutes) plus the buffer (240 minutes) plus the run time for all but the first part A (80 minutes), or 320 minutes (see Fig. 27).

Resource R-4

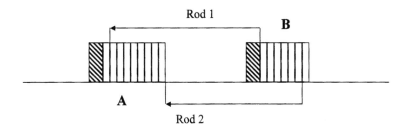

Figure 26   Backward rods.

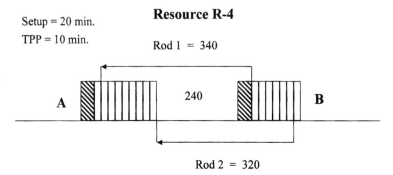

Figure 27   Computing the length of backward rods.

Rod 2 equals the run time for all but the last B (60 minutes) plus B's setup time (20 minutes) plus the buffer (240 minutes), or 320 minutes.

4. Combination Rods

Combination rods extend both forward and backward. The lengths of the rods are computed the same as in Figs. 25 and 27 for each of the other rod types (see Fig. 28).

The objective of the combination rods is to ensure that if A were to move to the right, B and C would also move. If C were moved to the left, A and B would move. In a complex environment there may be many orders attached in sequence.

Resource R-4

Figure 28   Combination rods.

Resource R-1

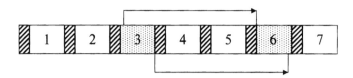

Figure 29   The impact of rods on setup savings.

5.   The Effect of Rods on Setup Savings

As explained, the purpose of rods is to maintain minimum distances between specific part/operations during the scheduling process. Part/operations 3 and 6 are scheduled on the constraint. Part/operation 3 is earlier and feeds 6. In Fig. 29 a decision has been made to perform setup savings by combining two similar part/operations with the same setup (2 and 4). Normally during setup savings the later part is moved forward and combined with the earlier part.

In Fig. 30 when bringing 4 forward, 3 was pushed into the future. However, that is also the direction of part/operation 6. Since 3 was pushed into the future, 6 was also pushed to maintain the minimum distance prescribed by the rods. Notice also that 7 took 6's position on the schedule.

6.   Recomputing Start and Stop Times

Figure 31 represents the constraint schedule and some detail information for R-4. F/30 and A/10 belong to sales order A for a quantity of 10 pieces. Since F/30 feeds A/10 and appears on the same schedule, rods have been created. Notice that F/30 has a pair of forward rods under the headings R-1 and R-2.

Forward rod 1 extends 340 minutes from the end of the run time of the first part F/30 toward part A/10. Forward rod 2 extends 385 minutes from the

Resource R-1

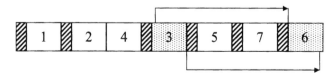

Figure 30   The impact of rods on setup savings—since 3 was pushed into the future, 6 had to be also.

Resource R-4

Sales Order/Qty	Part/ Oper.	Proc. Qty.	Setup Time	Run Time	Total Time	Start Time	End Time		R1	R2
S/O-A/10	F/30	10	10	10	110	98/3:27	98/5:17	F	340	385
	A/10	10	10	15	160	99/1:17	99/3:57	B	340	385
S/O-B/10	B/20	10	10	10	110	98/5:18	98/7:08			
S/O-C/20	C/30	20	10	10	210	99/3:58	99/7:28			
S/O-D/10	D/10	10	10	10	110	99/7:29	100/1:19			
S/O-E/20	E/30	20	10	10	210	100/1:20	100/4:50			

Figure 31 Recomputing start and stop times maintaining the rods. Constrain buffer is 480 minutes.

Resource R-4

Sales Order/Qty	Part/ Oper.	Proc. Qty.	Setup Time	Run Time	Total Time	Start Time	End Time		R1	R2
S/O-A/10	F/30	10	10	10	110	98/3:27	98/5:17	F	340	385
	A/10	10	10	15	160	99/2:40	99/5:20	B	340	385
S/O-B/10	B/20	10	10	10	110	98/5:18	98/7:08			
S/O-C/20	C/30	20	10	10	210	98/7:09	99/2:39			
S/O-D/10	D/10	10	10	10	110	99/5:21	99/7:11			
S/O-E/20	E/30	20	10	10	210	99/7:12	100/2:41			

Constrain Buffer = 480 minutes

Figure 32 Computing start and stop times maintaining the rods. Constrain buffer is 480 minutes.

end of F/30 to the beginning of the last part A/10. A/10 has a pair of rods which extend toward F/30.

During the scheduling process, which starts with F/30 and ends with E/30, the forward rods from F/30 were used to push A/10 into a different time slot. Rod 1 extends from 3:47 of day 98 (the time was computed as hour 3:27 plus setup of 10 minutes plus the 10-minute run time for one part) until 1:27 of day 99. Less the setup time for A/10 of 10 minutes, this places A/10's start time at 1:17.

Part/operation B/20 was scheduled to start immediately following F/30. Notice that C/30 is scheduled to start at 3:58, immediately following A/10. The scheduling of A/10 at the end of the rod has caused a 2 hour and 9 minute hole to develop in the schedule from day 98 at 7:08 to day 99 at 1:17.

The easiest way to solve the problem is to delay the scheduling of A/10 until immediately after C/30 on day 99 (see Fig. 32). A/10, because of the rods, can be scheduled further away, but not closer to F/30.

IV. SUMMARY

It is the scheduling of the constraint which establishes the rate at which throughput enters the company. If the constraint is able to process more orders, then the amount of money the company makes is maximized. The exploitation phase is designed to enhance that effort and means a tremendous difference in how resources should be scheduled.

V. STUDY QUESTIONS

1. List at least five ways of increasing productive capacity on the constraint.
2. Using the output from Question 12 at the end of Chapter 6, recompute the start and stop times of all orders by adjusting for time zero.
3. Define the concept of rods, name each type, and show an example of how they are used.
4. How are orders in the schedule for the constraint recognized as having a problem with being delivered on time?
5. What are the rules involved with overtime assignment and what method is used to limit the amount of overtime applied?
6. Apply setup savings, recomputing the start and stop times, to the output of Question 1 and define the recommended method of limiting the amount of orders being combined within the horizon.
7. Explain the impact of rods on the setup savings process.

8
Subordination

The objective of the subordination phase is to coordinate the activity of all other resources within the factory to ensure that the schedule established for the constraint is protected, that shipping is completed on time, and that raw material is released to the floor in synchronization with the way in which it will be consumed. It is an extremely important step in the process.

I. CHAPTER OBJECTIVES

- To define the subordination process
- To explain the methodology for the creation of secondary/tertiary schedules
- To present methods for resolving schedule conflict
- To define the process for scheduling the release of raw material

II. THE SUBORDINATION PROCESS

Once the schedule has been created by the exploitation process, it is the subordination phase which defines the activities of all other resources by ensuring that certain areas of the factory are provided with what they need (Fig. 1). It adds a dynamic quality to the system by performing the following functions:

- Determining whether there is enough protective capacity on the nonconstraints to protect the schedule and automatically increasing the size of the buffer for those buffer origins needing additional capacity

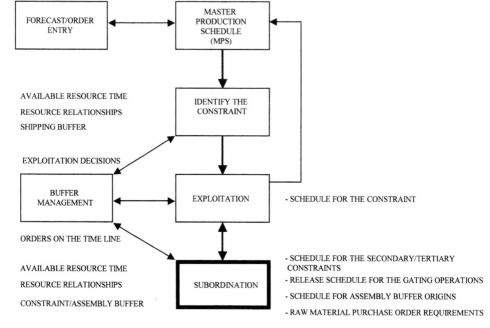

Figure 1 The TOC-based system—subordination.

- Identifying those resources for which protective capacity is inadequate (secondary/tertiary constraints) and which threaten the schedule of the primary constraint
- Exploiting those resources identified as additional physical constraints to maximize their ability to deliver to the primary constraint
- Ensuring that there are no conflicts between the schedule created on the primary constraint and the schedule created for the additional constraints
- Preparing the release schedule for raw material into the gating operations and synchronizing the activities of the nonconstraints

III. ESTABLISHING THE BUFFER SYSTEM

The heart of the subordination process is called the buffering system. Since it is the buffer that is used to guide the instructions given to the remaining resources, it literally defines how the subordination process is to be executed. Its role is to ensure that each buffer origin (the area of the factory being protected) is protected from disturbances in the system.

Buffers are established based on those physical laws which govern the manufacturing environment. There are three types. Each serves the purpose of protecting certain areas of the factory from those things which will inevitably happen and cause the creation of throughput to be threatened. Those buffers include

- The *shipping buffer* is used to protect the shipping of finished goods, to determine the initial schedule for the constraint, and to establish the release schedule for raw material which does not go through the constraint or assembly buffer.
- The *constraint buffer* is used to protect the constraint(s) and to determine the release of raw material to those operations which feed the constraint.
- The *assembly buffer* is used to ensure that those assembly operations directly fed by the constraint do not wait for material to arrive from nonconstrained legs within the net. It also determines the release schedule for raw material into those operations which feed the assembly buffer.

It is the buffering system which synchronizes the way material flows through the factory with the way in which it is consumed and determines the release schedule for raw material. In doing so it serves to protect the creation of throughput (see Fig. 2).

Each buffer is marked by a rectangular block which designates those resources which are included in the buffer process defined in bold print. Notice that the shipping buffer protects those resources that feed shipping and

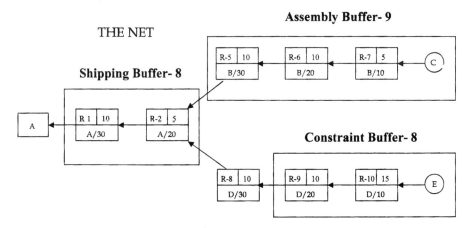

Figure 2 Establishing the buffer system.

in this case is 8 hours long. The constraint buffer protects resource R-8, which has been determined to be the constraint and is also 8 hours long. Finally, the assembly buffer protects R-2 and is 9 hours long.

The release date for raw material C is computed as the sales order due date minus the shipping buffer minus the assembly buffer. If A is due on day 100 at hour 8:00 (assuming an 8 hour day), then C should be released at day 98 hour 7:00. The due date for raw material E is the start date of the order across the constraint minus the constraint buffer. Raw material E should be released to the floor at day 98 hour 8:00. For those raw material items which do not go through the constraint or assembly buffers, the ship date minus the shipping buffer is used.

Material is scheduled in the buffer system by the demand placed on the buffer origin. If a sales order is due by a specific date, then the due date establishes the demand placed on the buffer. The schedule of the constraint causes its demand to be executed within its buffer. The same rule applies to the assembly buffer.

The actual length of the buffer will vary depending on the environment and characteristics of the resources involved. Buffers shrink or grow based on the output of the buffer management system. If material arriving at the buffer origin is consistently late, then the buffer is increased. If it arrives consistently early, then the buffer size is decreased.

A. Buffering the Constraint

It is very important to ensure that the constraint, like shipping, is protected from those things which can and will go wrong. The constraint buffer synchronizes that portion of the factory existing in front of the constraint with the constraints schedule. It also determines the release schedule for the gating operations which will eventually feed the constraint (see Fig. 3).

The sales order is due on day 102 at hour 8:00. There is a 16-hour buffer between the due date of the sales order and the completion of part/operation B/30 on the constraint. So B/30 must be completed by day 100 hour 8:00. Setup and processing require 2 hours. Consequently, B/30 must start at day 100 hour 6:00. The amount of protection for the constraint is estimated to be 16 hours as well. The release of raw material to the gating operation comes at day 98 hour 6:00.

Because of the constraint buffer, raw material will be released at a specific distance in front of the constraint to ensure that, despite any unforeseen problems, the parts will reach the constraint before the constraint is scheduled to consume them.

Remember that the schedule for the constraint may modify the release schedule for raw material. Had D/30 been scheduled at an earlier

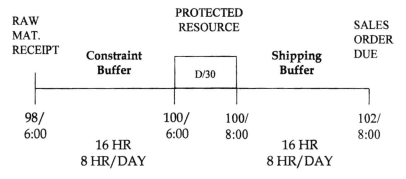

Figure 3 Buffering the constraint.

time during the exploitation process, then the release of raw material would have been offset by the constraint buffer from the earlier start date and time.

B. Assembly Buffers

Assembly buffers are used on those sections of the net that may not be constrained but feed assembly operations which are. They work exactly as constraint or shipping buffers. Notice that the scheduled start date of A/20 from Fig. 4 is used to offset for the release of raw material. Part/ operation A/20 is an assembly operation with at least one leg being fed by the constraint.

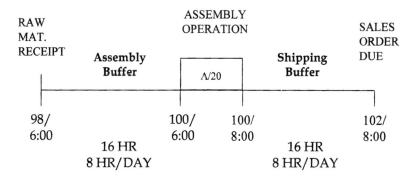

Figure 4 The assembly buffer.

IV. DYNAMIC BUFFERING

The heart of the subordination process for the information system is *dynamic buffering*. Dynamic buffering is a process which increases the size of the buffer whenever a lack of adequate protective capacity is detected by the system. If a resource load threatens the ability to deliver to the constraint on time, then the release of raw material is automatically pushed into an earlier time period. It is this feature which adds greatly to the effectiveness of the manual drum-buffer-rope (DBR) process described in *Synchronous Manufacturing* (1990) and in *The Theory of Constraints: Applications in Quality and Manufacturing* (1997).

Buffer requirements can be separated into fixed and variable types. The fixed buffer is that portion of the system's buffer which will handle most of the disturbances created by the variation in the ability of resources to perform. The variable buffer is that portion of the buffer which increases or decreases depending on the impact of demand on the schedule. If the demand for a given period of time is greater than the capacity of a resource can handle, it pushes the demand into an earlier period. Pushed far enough, it extends the size of the buffer.

The advent of dynamic buffering has added a great deal to the entire buffering process. By increasing the size of the buffer in only those places and times where it is absolutely necessary, the size of the fixed portion of the buffer can be reduced considerably. Since it is the buffer which determines the release time for raw material, there is a direct effect on the amount of inventory now required to offset for disturbances in the system. Rather than having a large buffer and the corresponding inventory which supports it, inventory will only be increased in those areas where additional protection is absolutely required. The size of the fixed portion of the buffer can be reduced.

A. Estimating the Load on the System

The first step in dynamic buffering is to develop an estimation of the load on the nonconstraints given the schedule for the primary constraint plus the demands created by the master production schedule (MPS). To do this means to first allocate inventory which already exists in the system and then compute the load. Allocating inventory in the subordination process, as in identification, is done on a first come, first serve basis. Since inventory availability has already been determined between shipping and the constraint, the allocation process must focus on those resources which occur after the constraint (see Fig. 5). The ten parts at part/operations H/20 and G/10 must be allocated to orders during processing.

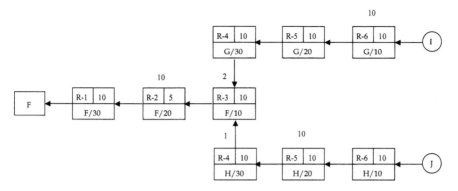

Figure 5 Allocating inventory.

Once the allocation of inventory has been completed, an estimation is made on the amount of capacity required by the nonconstraints in order to meet the demand created at the buffer origins. This task is done in much the same fashion as the initial placement of the load for identifying the constraint. The actual resource demand is placed at the end of the buffer and is not leveled. The initial task is to compare capacity with demand as in traditional capacity requirements planning (CRP) (see Fig. 6). That resource which has, on average, the highest demand compared to capacity requires additional scrutiny.

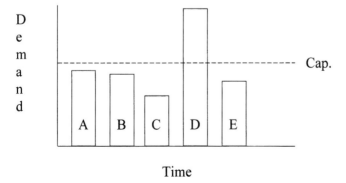

Figure 6 Capacity requirements planning.

Resource R-4

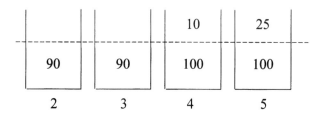

Figure 7 Resolving the peak load.

B. Resolving Peak Load Problems

Nonconstraints by definition have enough capacity, on average, to manage the demand which has been placed on them. However, there will be times when that demand is temporarily greater than the available capacity (see Fig. 7).

The fixed buffer for the constraint places the release date of raw material at the beginning of period 2. However, resource R-5 suffers from a temporary overload in periods 4 and 5. Resource R-5's capacity is placed at 100 hours per day. Period 5 has 25 hours of demand which cannot be met on that day, while period 4 has 10 hours. Periods 2 and 3 each have 90 hours of demand so additional time is available for those periods.

As seen earlier, if the demand on a resource exceeds capacity, the only direction in which to place the load is to an earlier time period. When this occurs on an occasional basis the fixed buffer is capable of managing the impact with little negative effect. However, if there is enough of a load, it can force the normal release time of raw material to be earlier than originally

Resource R-5

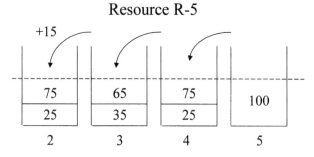

Figure 8 Dynamic buffering—an excess of 15 minutes remains.

Resource R-5

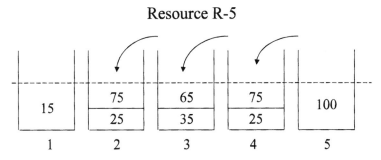

Figure 9 Dynamic buffering—excess 15 minutes is placed in period 1.

planned. In Fig. 8 there is an additional 15 minutes which cannot fit into period 2.

The fixed buffer is placed at the end of period 2, but additional time is required to ensure that resource R-5 can meet the demand of the constraint. So the additional demand is placed in period 1 (see Fig. 9).

If the demand on the nonconstraint is large enough, it will cause the release of raw material to occur earlier. Notice that 15 minutes of the demand was pushed into period 1. This means that the buffer for this portion of the net was enlarged by one period.

C. Dealing with Prolonged Periods at Near-Capacity Levels

Whenever a resource has demand placed on it which exceeds the estimated protective capacity required to successfully meet the demands of the buffer origin, it may be necessary to introduce periods of artificial load which will take the place of real demand and force the buffer to be enlarged (see Fig. 10).

Resource R-5

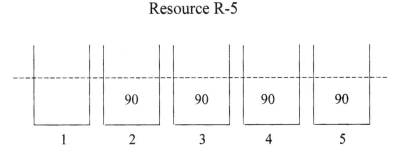

Figure 10 Dealing with prolonged demand at near-capacity levels.

Resource R-5

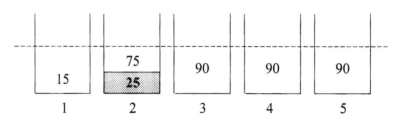

Figure 11   Applying artificial demand.

Resource R-5 has been loaded to near-constraint levels for four consecutive periods. This means that for those periods whenever a problem arrives there is not enough protective capacity available on a temporary basis to deal with the problem. In Fig. 11 an artificial load was placed on resource R-5 in period 2 which, in turn, increased the size of the buffer even though R-5 was never loaded over 100%.

D.   Identifying the Secondary Constraint

The secondary constraint is that resource which has created the greatest threat to the protection provided by the system (the buffers). In the preceding example, if resource R-5's demand were to be pushed so that it accumulated excess demand against time zero or excess load against an order which has already been scheduled on the constraint, then it is said to have created a peak load in period 1. While the largest peak load does not necessarily identify the

Resource R-5

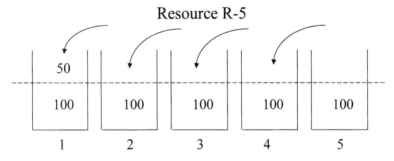

Figure 12   Peak load in period 1.

secondary constraint, it is one among several clues which will help to identify the next problem resource.

In Fig. 12 the load is being steadily pushed from period 5 toward period 1. Since period 1 begins at time zero the load can no longer be pushed, and it begins to accumulate there. Notice that period 1 has 50% more load than it can handle.

When examined from a global perspective, the secondary constraint is that resource which has done the most damage to the protection needed for the system to perform to current demand. So to understand the impact of a peak load in the first period it must be compared in relationship to the protection (buffer) being used.

Resource R-1 in Fig. 13 has an overload in period 1 of 50%. Resource R-2 has an overload of 100%. The buffer sizes (protection) for both resources are exactly the same. This means that the relationship of the amount of overload to the amount of protection is much different for the two situations. Resource R-2 will have a more difficult time spreading the larger peak load among the various resources within the buffer than will resource R-1.

In Fig. 14 the size of the buffer has changed. Notice that the first period peak load for resource R-1 happens within a buffer whose size is 50% smaller. It has less opportunity to spread the load, which means that when setup savings or overtime are considered to solve the peak load problem, resource R-1 will have a greater problem in resolving the situation. It is therefore possible that R-1 may be the secondary constraint. A good measure for identifying the secondary constraint is the ratio between the buffer size and the overload.

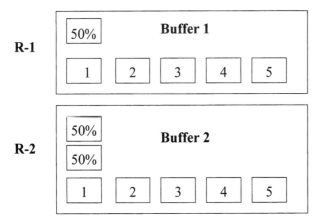

Figure 13 The relationship of the load to the size of the buffer.

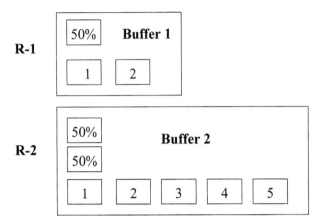

Figure 14 The relationship of the load to the size of the buffer—buffer 1 has been reduced.

If the amount of overload in minutes for period 1 exceeds the length of the buffer by a given amount, it is considered a problematic resource. That resource whose first period peak load exceeds the buffer size by a given amount greater than any other is identified as the secondary constraint.

E. Dealing with Peak Loads in the Red Lane

The *red lane* is defined as that area of the net which occurs between shipping and the primary constraint. It is particularly important because once the schedule has been set for the constraint and a large enough peak load is created on a resource existing in this area, it can threaten the schedule created by the constraint or result in pushing the sales order due date beyond its current schedule. The peak load problem for the red lane is handled the same way that a peak load is managed for those which occur at period 1. Whenever those resources are identified as being problematic (where the average demand exceeds the average capacity), the load is pushed toward time zero. However, in this case the load will be blocked from reaching time zero by an order or part/operation which has already been scheduled on the constraint. This order cannot be moved without damaging the constraint's schedule (see Fig. 15).

Resource R-5 was scheduled past 100% on days 24 and 25 and is pushing the load toward time zero. Part/operation A/20, which is fed by part/operation A/30 from the constraint schedule, was selected to be pushed into

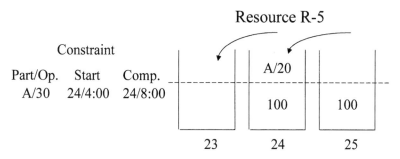

Figure 15 Peak loads in the red lane.

period 23. If pushed into an earlier period, it would have been scheduled at day 23 hour 0:01. Notice that part/operation A/30 was scheduled to start on the constraint at day 24 hour 4:00. Since A/30 feeds A/20 it would be impossible to have A/30 deliver to A/20 if A/20 is to start before A/30. Part/operation A/20's demand must accumulate on day 25. Whenever the demand accumulates on a specific resource beyond a certain ratio in comparison with the buffer, it is declared a problematic resource. That resource which is declared problematic and accumulates more than any other resource in the red lane is a secondary constraint.

F. Creating Secondary Schedules

Schedules for secondary constraints derive their priority from the buffer origin they are trying to protect. They can be required to feed either the primary constraint, the shipping schedule, the assembly buffer, or another secondary constraint.

The initial phase of creating the secondary constraint is much the same as in the initial phase of the primary schedule. The first step is to place the orders on the timeline in their ideal position based on the buffering system. The next step is to place the batches so that there are no conflicts with the existing schedule on the constraint.

Note: when interactive constraints exist only half the constraint buffer is used to set them apart (see Fig. 16).

Since D/20 feeds D/30 and D/20 is processed on the secondary constraint, then D/20's stop date and time will be one-half a constraint buffer width from the start time of D/30 on the primary constraint. If the constraint buffer were normally 8 hours and D/30 starts at day 100 hour 6:00, then D/20 should stop at day 100 hour 2:00.

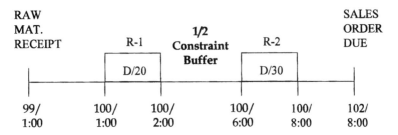

Figure 16   Interactive constraints.

Notice that D/20's buffer is a full constraint width. The gating operation is scheduled to release raw material at day 99 hour 1:00. Notice also that the protection in front of the primary constraint is now 50% larger because a secondary constraint was declared. Material is released 50% earlier. Whenever a secondary constraint is declared in front of the primary constraint in the system inventory will go up.

1.   Resolving Conflicts Between Schedules

The secondary constraint's schedule may look much the same as the primary schedule did as orders were initially placed on the timeline. However, with the secondary constraint there may be conflicts which arise when trying to place the load. It is advantageous for every order to be placed with enough time to ensure that the material will reach the buffer origin in time. As part/operations are moved in the process of exploitation, some orders will inevitably be placed closer to the schedule on the constraint than is desired.

Resolving conflict between schedules means ensuring that part/operations are protected in time. Like the rods discussed in Chapter 7 used to keep part/operations apart as they cross the constraint, rods are also used between schedules of two different constraining resources. Rod timing is computed exactly as in Chapter 7.

Figure 17 is an example of a forward rod used on multiple constraints. Resource R-4 is the primary constraint and is fed by R-3. Resource R-3 is the secondary constraint and is connected by forward rods 1 and 2, which originate from part/operation A and connect to part/operation B. During the scheduling process, the rods prevent part/operation A from approaching part/operation B by less than one-half the constraint buffer. Notice that these rods are connected between two different schedules. Since part/operation B was scheduled first on the primary constraint it does not move. Part/operation A must be scheduled around B. (Time runs from left

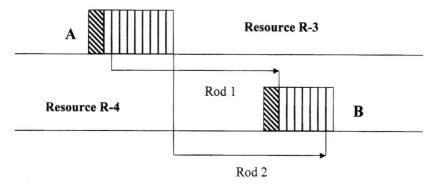

Figure 17 Forward rods on multiple constraints.

to right.) The objective here is to ensure that after part/operation A is processed on the secondary constraint it will have enough time to travel to the primary constraint. Resource R-4 is still the constraint, and part/ operation A still feeds part/operation B. However, A is now being processed on the primary constraint. As R-3's orders are being placed on the timeline, rods 1 and 2 prevent B from approaching A by less than one-half the constraint buffer. Figure 18 is an example of backward rods on multiple constraints.

Figure 19 is an example of combination rods between two resources. Resource R-4 is still the constraint. Part/operations A and C are those which have established schedules on the primary constraint. Part/operation B must try to fit between A and C without disturbing their schedules. It B must

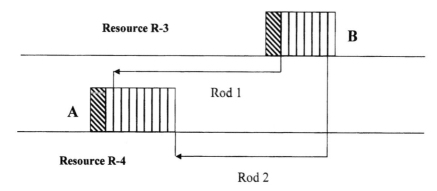

Figure 18 Backward rods on multiple constraints.

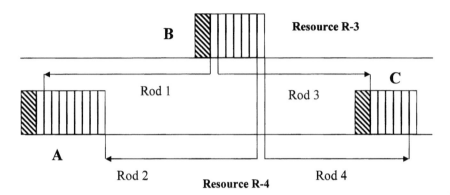

Figure 19 Combination rods on multiple constraints.

therefore be scheduled so that it does not violate rods in both directions. Because it must be placed between two established batches on the primary constraint, this is the most difficult batch to place in time and is given priority during the subordination phase.

Note: It is possible to have primary and secondary constraints which do not interact in any way. In this case rods are not required. Each part/ operation resource is scheduled independently of one another.

G. Placing the Load

The placement of the load for the initial schedule is according to the ideal location without consideration for the load or conflicts which may exist.

The first orders to be placed on the schedule for the secondary constraint are those that have combination rods. Part/operation 1 on the constraint feeds part/operation 6 with an intermediate operation on resource R-1. Resource R-1 is the secondary constraint. Part/operation D in R-1's schedule must be placed in time between 1 and 6 and be separated by combination rods equal to one-half the constraint buffer. Part/operation D is placed according to the backward time rod (see Fig. 20).

The next orders to be placed are those with forward time rods. The placement is one-half buffer constraint distance behind the connecting order on the constraint (see Fig. 21). Part/operation B from resource R-1 feeds part/ operation 4 on the constraint and has been placed one-half buffer length prior to part/operation 4. Part/operation E was placed one-half buffer length in front of 7.

Secondary Constraint

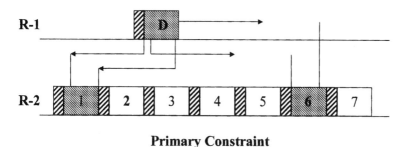

Primary Constraint

Figure 20   Placing the load according to the rods.

Notice that part/operations B and D are trying to occupy the same position on the schedule for R-1. When the final schedule is created, D will have priority in placement over B since D has combination rods and was placed first. Notice also that B has room to move to the left and not violate the forward time rods to 4.

The next placement is for orders with backward time rods. They are placed one-half constraint buffer distance in front of the connecting order on the constraint (see Fig. 22). Order F on resource R-1 is being fed by order 3 on

Secondary Constraint

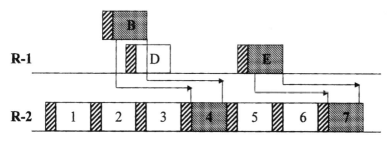

Primary Constraint

Figure 21   Placing the load according to the rods.

Secondary Constraint

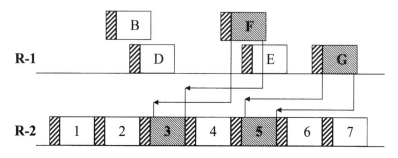

Primary Constraint

Figure 22 Placing the load according to the rods.

resource R-2 and was placed one-half constraint buffer width behind order 3. Order G is connected to order 5.

Notice that E and F are also trying to occupy the same space on the schedule for R-1. In this case priority will go to E since it has already been placed. Due to the backward rods to 3, F can only move forward in time to be placed.

Orders with connecting orders on the secondary constraint are placed one-half buffer distance behind the order which it feeds (see Fig. 23). Part/

Secondary Constraint

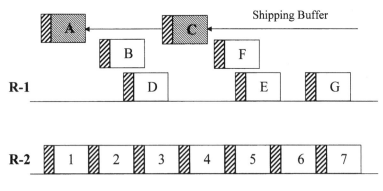

Primary Constraint

Figure 23 Placing the load according to the buffer.

operation A feeds C on the same schedule so it is placed one-half buffer width in front of C. Part/operation C is scheduled based on the shipping buffer.

Finally, the initial placement of orders which do not have connecting orders on any constraint are placed one shipping buffer from the due date of the sales order.

V. EXPLOITING THE SECONDARY CONSTRAINT

To exploit the secondary constraint means to schedule it so that there are no conflicts with the primary constraint and also so that the amount of available time is optimized to protect the constraint or any other buffer origin that the secondary constraint is feeding.

The first step in the process is to place the orders on the timeline based on the various buffer types. This was accomplished in the last section. The next phase is to level the load on the secondary constraint (R-1). Orders must be placed so that there are no conflicts with the constraint's schedule, the schedule for the secondary constraint, time zero, or the shipping buffer.

The first order to be placed during the initial placement on the time-line was D (see Fig. 24). During the leveling process, D's position is not going to change. The next orders to be placed are those with forward rods (see Fig. 25). Notice that the conflict between B and D has been solved. Since B could move to a position earlier in time it did not violate its forward time rods to 4 and no longer occupies the same place with D. Notice that E's position was not affected.

The next orders to be placed are those with backward rods (see Fig. 26). Orders F and G are pushed forward in time and placed immediately after F.

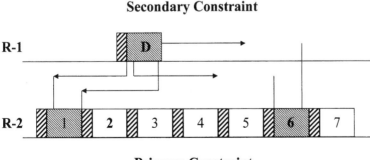

Secondary Constraint

Primary Constraint

Figure 24 Scheduling the secondary constraint.

Secondary Constraint

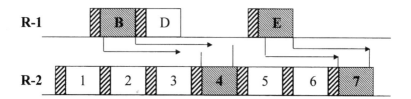

Primary Constraint

Figure 25 Scheduling the secondary constraint—placing orders with forward rods.

Notice the additional distance between F and G and their corresponding orders on R-2.

The last orders to be placed on R-1 are those which do not have any connection to R-2 (see Fig. 27). Order A feeds B on resource R-1 and is scheduled one-half buffer width in front of C. The arrows for the shipping and one-half constraint buffer have been placed in relative position. Notice that A has been moved to the left a short distance and placed in front of B. Notice also that C was moved a short distance toward the right and into the shipping buffer. Fortunately the schedule did not require more than a small portion of the protection of C to resolve all conflicts between the schedules for R-1 and

Secondary Constraint

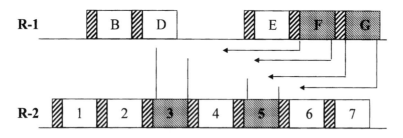

Primary Constraint

Figure 26 Scheduling the secondary constraint—placing orders with backward rods.

Secondary Constraint

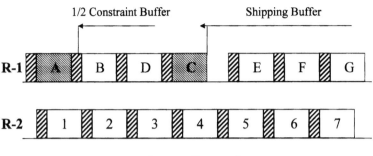

Figure 27 Scheduling the secondary constraint.

R-2. Had C been pushed farther to the right so that it took more than half the shipping buffer, the sales order which C feeds would be considered late and pushed into the future. Notice that there is a gap between orders C and E. In order to solve the conflict between the schedules, no orders are allowed to occupy this space.

The objective in exploiting R-1 is not to maximize its production. Once all conflicts have been resolved, the scheduling process stops. Had there not been enough capacity on R-1 to resolve all the conflicts and handle the load, issues such as setup savings would be considered. In this case it was not needed.

A. The Timing and Direction of Order Processing

In placing the load and scheduling R-1 there was a certain priority in the sequence that the orders were placed. One issue not discussed was the direction in which processing takes place:

- When placing the orders with combination rods, since each order is placed at the end of the backward rod, the process begins with the earliest order and moves to the latest.
- When placing orders with forward rods the timing of placement moves from the latest orders to the earliest.

Once the orders with combination and forward rods have been placed, an adjustment should be made for time zero. All the orders which have been placed past time zero should be pushed into the present. The resulting schedule should reveal any conflicts which exist in the two schedules

- *If no conflict exists*: The orders with backward rods are placed moving from earliest to latest.
- *If there is still no conflict*: Those orders with no rods are placed from the latest to the earliest.

Note: If a conflict arises between the primary and secondary constraint which cannot be resolved during the scheduling process, orders on the constraint are pushed into a later time period, resulting in sales orders being scheduled further out in time. Once this occurs, the scheduling process of the primary constraint must begin again with the new schedule date for the sales orders as input.

VI. PROCESSING ON THE NET

Unlike material requirements planning (MRP), which uses a concept called low level coding to determine the order of processing, the TOC-based system uses time. In low level coding MRP waits until that level in the product structure has been reached where all of the parent requirements and timing for a specific part have been computed before processing the gross-to-net requirements (see Fig. 28).

Part C appears on three different levels in the product structure. Before gross-to-net requirement generation is performed all of the planned releases for the parent items for C must have already been computed down to and including level 4.

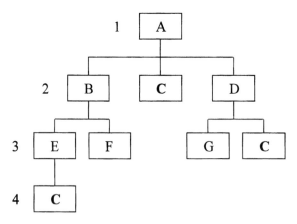

Figure 28   Low level coding.

In the TOC system, time is used. Since the system works with time as the controlling factor, knowing when something is to occur is the key issue. There are two places in time that are known to the system when the subordination process starts. One is the time when the sales orders are due, and the other is the schedule for the primary constraint. Thus, when moving in time these two issues must be considered.

The system starts with a block of time at the end of the schedule and determines what activity must take place during that time block. The next time block to be considered is the one which exists toward time zero from the end of the horizon. All activity for the first time period must be concluded before moving to the next.

Activity for a certain time block may be nothing more than computing when to place a specific order's part/operation in time so that when the system reaches that date while moving toward time zero it can place the load on the associated resource.

The system starts with a known occurrence at the end of the planning horizon. This will be a sales order. A list is used to keep track of those items which are identified in the current time as needing some sort of activity but for which action is to take place at a later time.

In Fig. 29 the system starts with the end of the planning horizon and moves in time constantly toward the present. Actions which are to be performed within certain time periods are done as the specific timing of that action reaches what is considered by the system as current time. Current time is the specific block of time that the system is working in as it moves from the end of the planning horizon to the present. As an example, when the system is computing requirements for day 200, then day 200 is the current day.

Before moving toward day 199, the system must complete all of the tasks that are assigned to be completed in day 200. Notice that in day 200 there is a sales orders for a quantity of 20 As.

One of the issues facing the system on day 200 is what must be done with the part/operations which feed shipping for sales order S/O111. Notice that

197	198	199	200

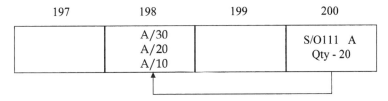

Figure 29   Moving in time.

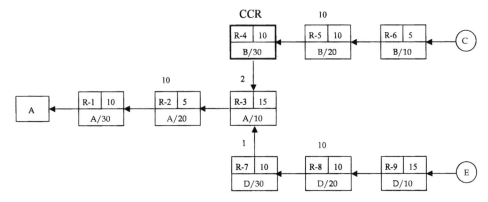

Figure 30   Processing on the net.

Part/operations A/30, A/20, and A/10 appear on a list of things to do on day 198. In this case the shipping buffer is 2 days long and, based on the requirements of the sales order, the system made a decision to place these part/operations on a list for day 198.

Figure 30 is the net that these data came from. The question which should be raised at this time is why stop at A/10? Why weren't part/operations B/30 or D/30 placed in day 198?

The objective in processing on the net is to find those things which have no place in time and to give them one relative to their position within the net. Part/operation B/30 for sales order S/O111 is processed on the primary constraint. It already has its place in time. Part/operation D/30 must be scheduled based on the assembly buffer established at A/10, so it would occupy a different day. If the assembly buffer for A/10 were one day, then D/30, D/20, and D/10 would occupy day 197 (see Fig. 31).

Once the processing down the net reaches a raw material part, the processing stops. The system returns to the highest assembly operation

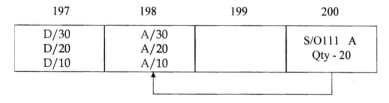

Figure 31   Moving in time.

and processes down the other leg placing part/operations on the list of things to do when the system's current date and the date of the list match.

Notice that after processing down the bottom leg in Fig. 30, when an attempt is made to process down the top leg, the first part/operation after the assembly is the constraint. Since the constraint has its place in time already, the processing stops and the next sales order is processed.

All sales orders for period 200 have been processed, so the current day moves to day 199. No orders exist in day 199 so the system again shifts the current day and it becomes day 198. Day 198 contains three part/operations originally placed by sales order S/O111, and each is to be processed on different resources. At this time the required setup and run times for each part/operation are placed on their respective resources. If an overload is detected by the system, the excess time will be pushed into an earlier time period, as in Fig. 9.

If additional sales orders existed in period 198, they would have been processed also. However, since there were none the processing for day 198 is complete, and the current day shifts again to day 197. On day 197 three part/operations are to be placed on their respective resources.

The next question that might be asked is why wasn't the inventory at operation A/20 considered when placing A/20 and A/30 in day 198. Remember that the process starts at the end of the horizon, not the beginning. That inventory has already been allocated to another order earlier in time. The first pass in processing on the net is to start from the present and move toward the end of the planning horizon allocating inventory.

The next step would be to place B/20 and B/10. When the system date reaches the date that B/30 crosses R-4 on the constraint schedule, B/20 and B/30 will be placed one constraint buffer in time in front of B/30 for sales order S/O111.

VII. SETTING THE SCHEDULE FOR DIVERGING OPERATIONS

The last subject to be discussed is the scheduling of diverging operations. Whenever one operation feeds two operations, the opportunity for the misallocation of material is very real. Material which is allocated to one part/operation on a specific resource may be taken by another. The usual reason is to promote the maximal utilization of one resource over another (see Fig. 32).

Resource R-5 has 15 parts which have been completed at part/operation G/30. The questions which must be answered are how many parts go to resource R-2 for processing E/10 and how many go to R-4 for processing F/

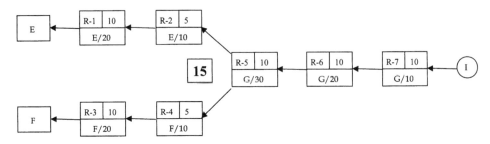

Figure 32   Diverging operations.

10. If R-2 were processing another part when time became available on R-4, R-4 would need to begin processing the G/30 parts. But how many should R-4 process? If R-4 were to stop at five parts and R-2 were still busy, there would be a tendency for R-4 to continue processing until all the G/30s were gone. Some of the G/30s would be needed to process the sales demand for E. However, there are no more G/30s left at resource R-5. They have all gone to be processed on those resources which feed F. To solve this problem a schedule must be created for R-2 and R-4. The key issue being how many parts to process and when.

While processing on the net, special treatment is given to those resources which process material arriving from diverging operations. A schedule is created consisting of a simple list of what parts must be processed that are arriving from diverging part/operations and which resource gets what quantity and when. Of the 15 parts in Fig. 32, five may be scheduled for order E and ten may be scheduled for order F.

VIII.   SUMMARY

The subordination process is important in that it defines how the remaining resources are subordinated to the demands of the schedule created for the primary constraint. The key issues are

- Providing the constraint what it needs to perform to the schedule created during the subordination phase
- Exploiting of secondary constraints
- Overcoming conflicts which may exist between the primary and any secondary constraints
- Processing on the net
- Buffering the remaining system

Subordination is an integral part of the five-step process of improvement and as such must be included in the construction of the information system.

IX. STUDY QUESTIONS

1. Define dynamic buffering and explain its overall importance to the scheduling process.
2. Explain the difference between the fixed and variable portions of the buffer.
3. What is the rule of thumb when establishing the initial buffer length to use when implementing a manual drum-buffer-rope process?
4. What is dynamic buffering and what is its impact on subordination?
5. What problems could be encountered between the primary and secondary constraints in creating a valid schedule?
6. Explain the sequence and direction of movement when placing orders on the timeline with specific rod types.
7. Explain the process of moving in time on the net.
8. Define the concepts of diverging and converging operations; explain at what time each may need a schedule; and provide insight into the kinds of problems which are caused by the overutilization of resources when confronted with these two characteristics.
9. Define the term *resource activation*.
10. What provisions have been made for dealing with prolonged periods at near-capacity levels on nonconstrained resources?
11. What is meant by the concept of a peak load on day 1, or in the red lane, and how does the system deal with these problems?

9
Buffer Management

The buffer is used to set the amount of protection needed by the system during the scheduling process and to set the raw materials schedule. However, it is not enough just to set the schedule. It must be controlled so that those things which are supposed to happen do happen. Buffer management represents the control portion of the system. It is used to ensure that if something goes wrong, it can be detected and fixed before there is an impact which will threaten the creation of throughput. There are two issues:

1. To identify those orders which are having problems and rectify the situation so that the orders will not be late
2. To identify and fix those things which consistently make orders late

I. CHAPTER OBJECTIVES

- To present the buffer management concept
- To understand the relationship between buffer management and the improvement process
- To present methods of using the system to accomplish buffer management

Buffer management is an important part of the Theory of Constraints (TOC)–based system. As seen in Fig. 1, it plays a part in identifying and exploiting the constraint as well as establishing the subordination process. Used effectively, buffer management

- Helps to identify and correct those orders which are having problems in reaching the buffer origin before they can affect the generation of throughput
- Establishes the focal mechanism for ensuring that when an action is taken to improve the amount of protection available in the

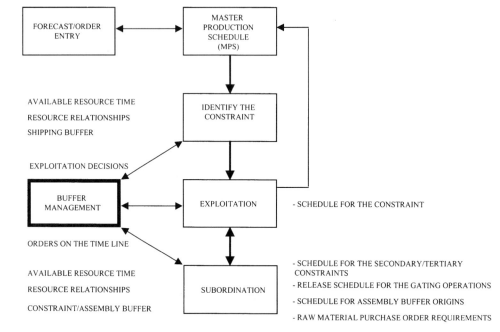

Figure 1 The TOC-based system—buffer management.

system that there will be an immediate impact on the reduction of inventory and operating expense
- Prioritizes problems in order of relative importance so that those issues having the greatest impact can be addressed first
- Aids in establishing the length of the fixed portion of the buffer

II. IDENTIFYING PROBLEM ORDERS

One of the objectives in buffer management is to identify problem orders before they impact the buffer origin. Problem orders are those orders which will threaten the creation of throughput. If they can be detected and action is taken early enough, then the buffer origin is adequately protected.

In Fig. 2, order A must travel from the gating operation, beginning at 6:00, through all the intermediate operations and arrive at the capacity-constrained resource (CCR) by 12:00. To arrive later than 12:00 means to lose throughput by not having an order to process on the CCR. However, it does not help to look at the order at 12:00 and determine that it

Figure 2 Identifying problem orders.

has not arrived; a loss in throughput will probably be unavoidable. To effectively protect the constraint, the time to examine A is before it is scheduled to arrive.

In Fig. 3 the time in front of the constraint on which A is to be processed has been divided into three different zones. Notice that zone I is from 10:00 until 12:00 and is called the expedite zone. If A does not arrive at the constraint before 10:00, it is located and expedited so that it will reach the constraint at least by 12:00, when it is to be processed. Zone II is from 8:00 until 10:00 and is called the tracking zone. If A does not arrive at the constraint before 8:00, it is located and tracked to see if problems might be developing. Zone III is considered normal time and runs from 6:00 when the order is released until 8:00 when the tracking zone begins.

A. The Zone Profile

Not only is it possible to identify those orders that are getting into trouble and expedite them, but also it is possible to look at a cross-section of how well the system is performing based on the percentage of those orders that either do or do not arrive at the constraint within a particular zone. In Fig. 4, 90% of all orders are sitting in front of the constraint within zone I. Seventy-five percent are arriving within zone II.

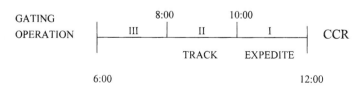

Figure 3 The buffer zones.

Figure 4   The zone cross-section.

This cross-section means that for the CCR in Fig. 4, 10% of the orders are being expedited.

A lower percentage of on-time arrivals could mean that the buffer is too small. The cross-section in Fig. 5 means that 25% of all orders are being expedited. If parts are being released too late, then the orders would arrive late. In order to solve this problem either the buffer should be lengthened or the cause of the late orders discovered and fixed.

Figure 6 represents the zone profile of a buffer origin as it appears at 1:00. A number of orders are approaching the buffer origin. Each order has its own place in time. Order number 1 starts at 1:00 and runs until approximately 1:25. Order 2 starts at 1:25 and runs until 1:45. The shaded areas with bold numbers indicate that the order has arrived at the buffer origin. At 1:00, zone I has all of its orders, zone II is short one order, and zone III has none.

Different zone profiles point to the occurrence of specific problems. Figure 7 shows another buffer. Notice that at 1:00 all orders except number 18 have arrived for all the zones. This is a good indication that the fixed portion of the buffer is too long and should be shortened. Figure 8 indicates that the buffer may be too short. Only half of zone I is filled. The numbers on the left represent time in 15-minute intervals. If this picture of the buffer is correct, then the buffer manager has 1 hour to find more work for the constraint or it will be idle.

The situation in Fig. 9 is much better than in Fig. 8. However, notice that two orders are missing from zone I, while other orders have arrived at zone II. Since there are other orders available in front of the constraint, this situation is not as bad. There is approximately 3 hours 10 minutes of

Figure 5   The zone cross-section—buffer too small.

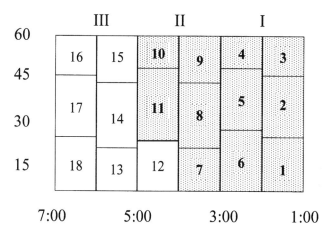

Figure 6 Buffer management—zone profile of buffer origin at 1:00.

work available. However, it does indicate that something must be wrong with the orders that were supposed to appear in positions 4 and 6. They should be found and expedited and the cause listed on the buffer management worksheet.

Figure 10 shows orders which have arrived at the constraint but are not inside any of the buffer management zones. Orders which arrive too early are an indication that they were released before their scheduled

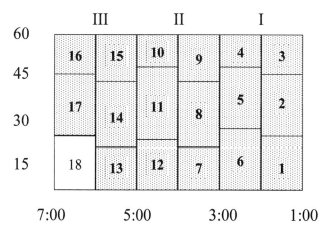

Figure 7 Zone profile at 1:00 showing buffer too long.

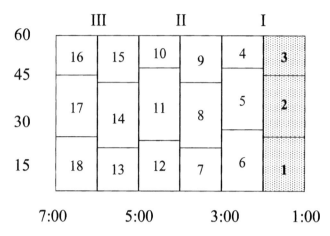

Figure 8 Zone profile at 1:00 showing buffer too short.

release date. This is usually caused by workers or supervisors who have a problem understanding that sometimes resources are idle for a reason. However, because the sequencing of work at nonconstraint operations is determined by the release schedule at gating operations, having an order arrive early at the constraint may mean that a temporary bottleneck will be created. The early order might have taken a specific resource time slot that was meant for another order.

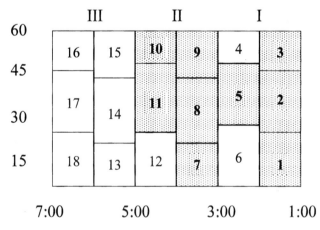

Figure 9 Zone profile at 1:00 showing holes in the buffer.

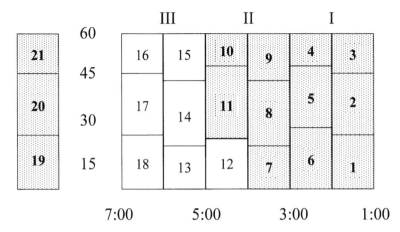

Figure 10 Zone profile at 1:00 showing orders released too early.

III. THE BUFFER MANAGEMENT WORKSHEET

Whenever the cause of an order's lateness in arriving at zone I is detected, it is placed on the buffer management worksheet maintained by the buffer manager. The list is consolidated and used to prioritize the order in which problems are fixed. Figure 11 is taken from *The Theory of Constraints: Applications in Quality and Manufacturing* (1997) and represents a buffer management worksheet.

This kind of worksheet can easily be placed into the shop floor control portion of any system for collecting data concerning the health of the system. As material arrives at the buffer origin it is logged, and an automatic percentage rating for each zone of the buffer can be calculated.

BUFFER ORIGIN: <u>RESOURCE R-4</u>   DATE: <u>09/02/95</u>

BUFFER LENGTH: <u>16 HRS</u>   ZONE DISTRIBUTION: I <u> 90% </u> II <u> 75% </u>

LOC	STATION	CAUSE	ZONE I DUE	ZONE I ACT.
R-1	A/30	MATERIAL SHORT	101/3:15	101/5:15
R-1	A/20	BROKEN FIXTURE	101/5:30	101/5:45
R-2	A/30	OUT OF SPEC.	101/5:45	101/7:17

Figure 11 The buffer management worksheet.

BUFFER ORIGIN: RESOURCE B   DATE: 02/27/94

BUFFER LENGTH: 16 HRS   ZONE DISTRIBUTION: I   90%   II   75%

LOC	STATION	CAUSE	TOTAL TIME		T VALUE	T$D
R-1	123/30	REWORK FOR SIZE	.5	X	$1,000	$500
R-1	123/20	KEYWAY TOO SMALL	1.0	X	$2,000	$2,000
R-2	121/30	MATERIAL SHORT	2.0	X	$1,500	$3,000

Figure 12   The buffer management report.

A buffer management report can be generated from the input data similar to Fig. 12, also taken from *The Theory of Constraints: Applications in Quality and Manufacturing*. Notice that the "throughput dollar days" have been calculated for each cause so the effort to fix problems that cause holes in the buffer can be placed in proper perspective. Throughput dollar days (T$Ds) were calculated as the length of time that the order was late getting to the constraint in zone I multiplied by the amount of throughput that was threatened. At resource R-1, part/operation 123/30 was reworked for size, causing the order to be a half day late. Multiplied by the $1000 of throughput generated by the order, the total throughput dollar days for the offense is $500. Compared to R-1's problem, R-2's, which caused $3000 in throughput dollar days to accumulate, was a greater threat to the creation of throughput.

After collecting the data, a Pareto analysis is performed to determine the priority of effort spent to solve the problems (see Fig. 13). The number one priority seems to be the seals problem at resource R-2, which has accumulated $12,000 in throughput dollar days. Number two is the boot problem. So the premier effort to increase the protection in front of the buffer origin should be to fix the problems with the seals, and the next to fix the boots problem.

A.   Accumulating Throughput Dollar Days by Department

In addition to focusing efforts to ensure that the primary causes of orders being late are fixed, attention must also be given to making sure that orders are expedited effectively. Any order which becomes late will begin collecting throughput dollar days for the department where the order is currently

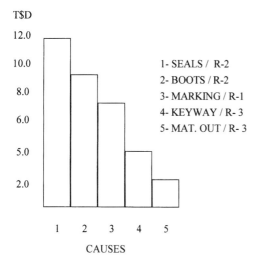

TSD

12.0
10.0
8.0
6.0
5.0
2.0

1- SEALS / R-2
2- BOOTS / R-2
3- MARKING / R-1
4- KEYWAY / R-3
5- MAT. OUT / R-3

1 2 3 4 5

CAUSES

Figure 13 Pareto analysis to determine priority of buffer problems.

located. The later it becomes or the higher the dollar value, the greater the pressure becomes to move the order to the next department. As soon as the late order leaves the department, the accumulated TSDs immediately drop for that department but increase for the succeeding department. This will continue until the order is shipped.

B. Inventory Dollar Days

There are certain actions which will have a negative effect on the system because they happen too early. As an example, if material is released to the gating operation earlier than the scheduled release date, inventory will increase. If material is processed too early at another operation, the result may be that parts in the production process will be mismatched. After the schedule has been created and instructions are given for the timing of specific events, the inventory dollar day (ISD) measurement is used to provide an incentive for preventing an activity from occurring too soon. It establishes the relative value of a specific activity. Here the value of the inventory increased by the action is multiplied by the number of days that are affected. As the value in the measurement increases, the pressure will increase to prevent the activity from occurring. As for throughput dollar days, a Pareto analysis of the inventory dollar days assigned to a specific cause or department will lead to an understanding of where to focus activity to make improvements.

IV. THE BUFFER MANAGER

Companies having implemented TOC in a manufacturing environment have found a need for a person to ensure that the buffer is maintained. This is a very important position. If the buffer is too long, this person is responsible to shorten it in the system, and vice versa. If there are holes in zone I of the buffer, orders must be expedited and data collected.

V. SUMMARY

Buffer management is a unique and powerful tool to increase the effectiveness of the scheduling system by controlling events and preventing them from negatively affecting the buffer origin. It is easily reproduced in the information system once the schedule has been created and will readily fit into shop floor control.

Used effectively the buffer management system can also help tremendously in the focusing process by identifying those actions which need to be fixed in the schedule to increase the amount of protection available for the buffer origin and to reduce inventory and operating expense.

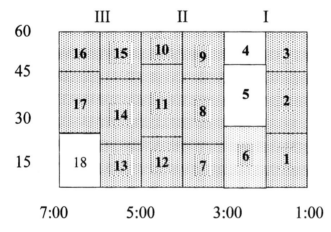

Figure 14 Buffer management example for study question 6.

VI. STUDY QUESTIONS

1. Define the concept of buffer management and describe the primary responsibilities of the buffer manager.
2. Explain what is meant by the *tracking* and *expedite zones.*
3. What is a zone profile and what is its primary function?
4. What are the purposes of the buffer management worksheet and the buffer management report?
5. Define and give examples of the use of the measurements *throughput dollar days* and *inventory dollar days.*
6. What situation(s) is indicated by the graphical representation of the buffer shown in Fig. 14?

10
Supporting the Decision Process

Once a complete definition of an information system has been made, how then should it be used? The system can provide information that will aid in the process of answering specific day-to-day questions such as what lot size should be used and what price should be accepted for this product.

I. CHAPTER OBJECTIVES

- To introduce the decision support features of the system
- To relate how the information system should be used to make decisions

The information system was designed to provide valid information, not just to schedule and control the factory. While a valid schedule and an effective control mechanism are prerequisites to the information system and are a tremendous breakthrough, it is the assistance in strategic as well as day-to-day decision-making which maximizes the system's effectiveness in helping the user to generate a profit. This portion of the book will be dedicated to using the information system to accomplish this task.

Note: A detailed discussion of the fundamentals of the decision-making process can be found in *The Theory of Constraints: Applications in Quality and Manufacturing* (1997).

Before continuing, it must be understood that the information system, like any other computer system, is merely a tool. Used incorrectly, the desired result will never come to fruition. Used correctly by intelligent and knowledgeable people, the information system can serve as a valid replacement for the current manufacturing schedule and control system as well as cost accounting. However, it will never run the company nor should it be considered infallible. Management's intuition about a particular issue in running the company is sometimes more accurate.

II. MAKING DECISIONS

Keep in mind that *information* is defined here as the answer to the
question asked. How each department within the organization should
use the system must be determined in the context of the decisions made.
It is the individual tasks to be performed which define the requirements.
Since the objective is to implement the five-step process of improvement,
the action required depends on where in the process the decision is
being made.

The discussion of how to use the system to make decisions will be
placed in the context of the two major categories of *tactical* and *strategic*
decisions.

- Tactical decisions are those decisions made on a day-to-day basis to
  run the company.
- Strategic decisions are those decisions that plot the direction of the
  company's future for years to come.

Tactical decisions include issues such as

- Where to focus improvements in the factory
- Whether to accept a specific offer for a sales order
- Whether a particular part should be purchased outside the company
  or made internally

These are the kinds of decisions that keep the wheels turning on a daily basis.
Strategic decisions include issues such as

- Market segmentation
- Company growth
- Long-term planning
- Recession-proofing the company

Strategic decisions are extremely important. A mistake here usually results in
long-term problems and even bankruptcy.

Note: strategic decisions will be discussed in Chapter 11.

A. Tactical Decisions

A complete discussion of how the information system should support the
decision process could not anticipate every decision that is to be made. The
objective here is not to present every possible situation, but to convey to the
reader the basic issues involved with using the system to make the day-to-day
decisions normally faced in the manufacturing environment. Tactical de-

cisions will be discussed in the succeeding sections on a departmental basis, as follows:

- Purchasing
- Sales and marketing
- Finance and accounting
- Quality assurance
- Engineering
- Production

III. PURCHASING

The general purchasing questions that must be answered are what to order, when to schedule delivery, and what to pay for raw material items. The TOC-based manufacturing system creates a list of raw material purchasing requirements similar to material requirements planning (MRP). In fact, the material requirements messaging facility for most MRP-based systems works very well in controlling the purchase of raw material for production. What the TOC-based system provides that MRP cannot is a valid schedule of requirements that closely resembles the way in which the material is consumed. Additionally, it provides feedback to the purchasing agent or manager regarding the impact or importance of a specific order.

Scheduling has proven to be very capable at handling the *what* and *when* questions. But how should the system regard the issues involving the cost of raw material? Since the goal of purchasing must be to protect the creation of throughput, the discussion should probably begin there.

Purchasing conflicts generally deal with the protection or the creation of throughput; common scenarios are

- The cost of a part is perceived to be too high in comparison to the individual profit margin obtained for the product.
- The vendor's perception of the value of the product is in conflict with the purchaser's.

So it seems that one advantage the information system could provide the purchasing department is to ensure that the purchasing agent understands the true value of the part being bought. This was the original intent of cost accounting. As discussed earlier, however, cost accounting long ago ceased providing a benefit.

The true value of a part is determined by the benefit derived. This brings up an interesting question. How can the purchasing agent sitting in an office perhaps hundreds of miles from the factory understand what benefit a specific part will have? Can the system help?

There are some key issues involved in understanding this problem. Besides the obvious raw material expense, there is a difference in the value of a part that depends on whether it is

- Processed on the constraint
- Using protective capacity
- Using excess capacity

A. Parts Processed on the Constraint

Sometimes it is better to determine the value of a part by looking at its absence. For instance, what is the value of not having a part that feeds the constraint? Is it different than for a nonconstraint? How will that impact the perception of the worth of the product to the purchasing agent?

In Fig. 1, raw materials A and B feed resources that eventually feed the constraint. In order to complete a sales order valued at $300, both A and B are needed. The normal price for A is $75; the standard cost is in the computer at $80. The vendor has some As located in another city and can have them air-freighted at double the price. B's cost is set at $100 and is available. Labor cost is an additional $55. The normal gross margin on this product is 23%. However, in this case the company is looking at a total cost of goods sold (COGS) of $305. From a cost accounting perspective spending the additional money to obtain the part is a losing proposition. The question is how can the system help solve this problem for the purchasing agent.

The first thing that can be eliminated from the equation is the $55 labor cost. Most companies pay based on a 40-hour week regardless of where the labor goes. That leaves the issue of whether the part should be flown in or

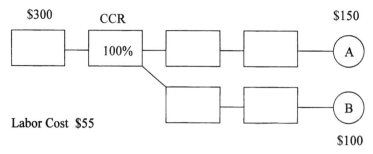

Figure 1 Making a purchasing decision.

delayed for a better price. If the part is not obtained and the constraint runs out of work, there will be an immediate $300 loss in total revenue being received by the company. There is a limited amount of capacity on the constraint that, once lost, can never be regained. If the part is obtained and the capacity-constrained resource (CCR) is able to continue working, the company makes $50 ($300–$250) and keeps the customer happy.

What information did the purchasing agent need to be able to make the decision?

- Information that part A fed the constraint
- Information on whether there is a hole in the buffer that cannot be recovered by processing another order
- The sales price of the end item
- The cost of raw material

If this is the information needed to make the decision, then obviously this is what the system must provide. If the constraint were unknown, the purchasing agent would first need to identify it. Clearly, the TOC-based manufacturing system is capable of accomplishing this task. However, in any company that has implemented TOC and has a physical constraint in production it is not kept a deep dark secret. Quite the opposite is true. So what purchasing needs is a method of identifying

- Orders which cause holes in the buffer
- The raw material parts required
- The sales price of the products that will be impacted
- The current cost of all raw material for the product/line item involved
- The projected due date of the purchase order line item in question

In Fig. 2, part/operation 123/30 of order number 2 in zone I of the buffer has not arrived at the constraint. This order is late because raw material part A has not yet been delivered by the vendor. This is not an acceptable situation. While additional orders are in zone I and can be used to fill the position of order 2, the probability is quite good that something else will occur that will threaten the constraint even further.

It's obvious that raw material A must be received very quickly. However, if the value of the sales order is $300, it makes no sense to pay $10,000 to get the part. So what should purchasing be willing to pay?

Figure 3 is a raw material pegging report screen for all raw material parts that feed the constraint. Included in the data are

- Raw material part
- Quantity of the raw material required

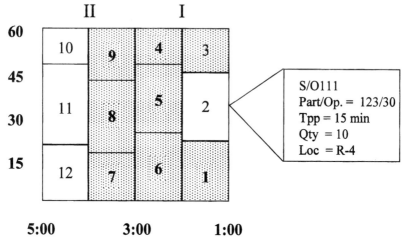

Figure 2 Buffer management—making a purchasing decision.

- Part/operation which is to be processed on the constraint
- Product involved
- Sales order
- Sales order value
- Total raw material cost included in the product

Notice that a quantity of ten As are required to feed part/operation 123/30. Notice also that sales order S/O111 is valued at $300 and has a total raw material content of $100.

This means that purchasing can pay up to an additional $199 for the ten part As that will be used on part/operation 123/30 to fill sales order S/O111

Raw Material Pegging/Constraint

Raw Mat.	Qty Req.	CCR Part/Op	Product	Sales Ord.	S/O Value	Total RM Cost
A	10	123/30	F	S/O111	$300	$100
A	20	124/20	G	S/O222	$150	$50
B	15	125/20	G	S/O222		
C.	10	127/30	H.	S/O333	$200	$100

Figure 3 Raw material pegging report.

and the company will still show a $1 profit. If there are not enough orders in the buffer to insure that the constraint continues to generate throughput and the raw material does not arrive in time, then the constraint will be idle and the company will lose $300.

Note: Whenever a sales order is threatened, the urgency of the customer's needs must also play a part. Sales must become involved. Ultimately it is a management decision.

1. Off-Loading Constraint Time to the Vendor

If all the demand for the constraint's time can be filled through exploitation without making orders late, then all parts should be produced in-house. However, if during the scheduling process all attempts to generate a schedule for the constraint fail to eliminate late orders, they can either be pushed into the future or additional resource time can be purchased. Purchasing additional constraint time includes off-loading constraint time to a vendor.

Note: Off-loading to a vendor raises similar issues to off-loading to an internal resource. The earlier in the process that off-loading takes place, the more orders will be impacted.

In making the decision, the amount of additional expense for purchasing resource time to process the part is added to the total cost of raw material for the product to determine the amount of throughput generated. If purchasing resource time will still generate a positive throughput figure, then the impact will be to increase total throughput (see Fig. 4).

The total sales of the company based on the schedule is $30,000 per week, while the total raw material required is $20,000. Product A has been sold for $300. The total cost of the raw material plus the expense of purchasing resource time from outside the company to make the part is $250. The additional throughput generated is $50. When this is added to the total throughput for the company, the resulting throughput is $10,050 per week.

		A			
Total Sales	$30,000	300			
Total RM	$20,000	250			
Throughput	$10,000	+ 50	=	10,050	

Figure 4 Off-loading to a vendor calculation.

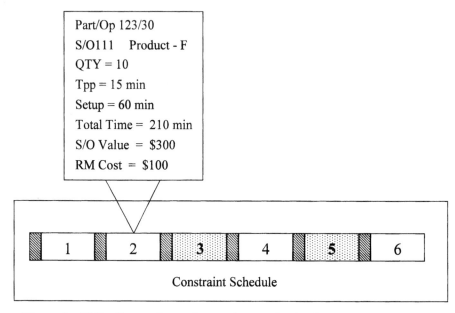

Figure 5   Off-loading to the vendor—orders on the timeline.

To make the decision the system needs to provide

- A view of the schedule including past due orders
- The part/operation to be unloaded
- The value of the sales order fed by the part/operation
- The total cost of the raw material involved

In Fig. 5, orders 3 and 5 are late even though all efforts have been exhausted to maximize the utilization of the constraint. Sales order S/O111 can be either pushed into a future period, thereby relieving the constraint, or part/operation 123/30 can be off-loaded to a vendor.

If a decision is made to off-load to a vendor, purchasing can spend up to $200 in additional expense before throughput will be impacted negatively. Up to $199 the company is generating more throughput.

B.   Off-Loading from Resources that Use Protective Capacity

One of the key issues in the previous discussion on the subordination process was the impact of conflicting orders. Whenever resources are loaded to the

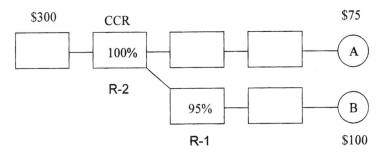

Figure 6 Estimating the value of a part.

point where they are unable to respond effectively to the demands of the constraint and additional schedules must be created on secondary constraints, there is always the possibility that a conflict will exist between orders on the primary and secondary constraints.

In Fig. 6, resource R-1 feeds resource R-2 and is loaded to 95%. Resource R-2 is the capacity-constrained resource and is loaded to 100%. Since R-1 feeds the CCR, its sole reason for existing is to provide R-2 with whatever it needs when it needs it. But R-1 does not have enough protective capacity to accomplish this task. Steps might be required to enhance its abilities by producing a schedule, combining setups, or applying overtime. To do this, R-1 could be declared as a secondary constraint and a schedule created. If this is the case, the chances are good that conflicts will exist between orders scheduled on R-2 and those scheduled on R-1. When this occurs and everything possible has been done to increase its ability to deliver to the CCR on time but conflicts still exist, there are really only three alternatives: postpone the sales order, off-load the time to other resources, or reduce the size of the rod or buffer which created the conflict.

When faced with this situation the system should identify all orders with rod or buffer conflicts and allow time for the decision to be made to either push the sales order into a later period, add temporary capacity, or off-load. It should also determine the impact of the off-load by continuing the subordination process with the remaining orders, as discussed in Chapter 8. If all of the current conflicts are resolved and no new ones are created, then the attempt was successful.

In some cases, off-loading to internal resources can cause additional problems. So a determination must be made whether to off-load demand from internal resources used to protect the constraint to an outside resource. How should this decision be made?

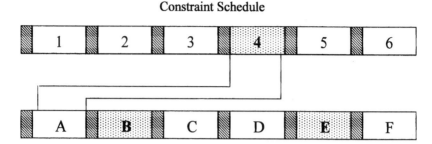

Figure 7   Constraint schedule showing rod violation.

Figure 7 indicates that part/operations B and E are late due to rod violations. Notice that part/operation 4 has backward rods that extend toward part/operation B. Notice also that part/operation B is in violation of those rods. The rods are there to ensure that part/operation B has enough time to finish being processed on the secondary constraint before being sent to the primary constraint. In this case there is insufficient protection between the completion time of B and the start time of 4.

As mentioned earlier, there are three possible choices:

- Push part/operation 4 into the future to the full extent of the rods and delay the sales order.
- Find additional capacity for the secondary constraint.
- Shorten the backward rods from order 4.

1.   Reducing the Size of the Rod

Rods can be reduced for selected orders when it is known that additional capacity is available, on a temporary basis, to help in the production of the conflicting order. Additional capacity can be added just prior to or when a specific order reaches the resource in question. In this case, another employee could be added to the secondary constraint for the processing of A and B and then removed when the problem has passed (see Fig. 8).

If additional capacity is not available to apply to the secondary constraint (whether for specific orders or in general) to solve the problem of conflicting orders, then perhaps outside capacity can help.

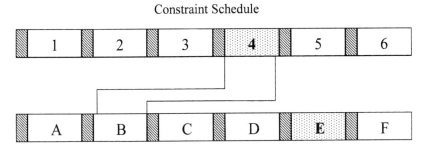

Figure 8 Constraint schedule showing reduction of rod size.

2. Off-Loading to the Vendor

If every possible alternative of finding additional internal capacity has been explored without success, then seeking an external source may be justified. There are two issues involved:

- The amount of throughput generated by the schedule when the order is delayed
- The amount of throughput generated when buying the processing time through an outside source

In Fig. 9, part/operation A has been off-loaded to a vendor, resulting in an elimination of the rod violations for B and E. Part/operation 4 is no

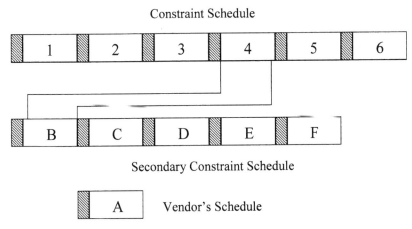

Figure 9 Constraint schedule showing rod violation.

Farm out cost		Total Raw Mat.		Total Variable Expense
$200	+	$2000	=	$2200

Sales Price		Total Exp.		Throughput
$3000	–	$2200	=	$800

Figure 10   Determining the amount of throughput.

longer threatened, and as a result the sales order will not be pushed into the future.

To determine the impact to throughput the additional expense associated with off-loading to the vendor is added to the raw material cost to determine the total variable cost of the order. The total variable cost is then subtracted from the sales order to determine the throughput generated (see Fig. 10).

3. Pushing the Order into the Future

When a sales order is pushed into the future due to a conflict between resources a new schedule is usually created for the constraint given the new due date for the sales order. When the constraint is rescheduled an attempt is again made to maximize its utilization. So the hole in the schedule created by pushing the new order into the future should no longer exist. Thus, the throughput generated by the constraint should not decline.

If the rescheduling process is not successful and a hole occurs in the schedule, throughput will decline (see Fig. 11). If the reason is due to a conflict with an earlier resource such as in Fig. 7, then the throughput generated using the additional expense of off-loading to the vendor needs to be compared to the estimated value of the unused CCR time. The estimated value of the unused CCR time is computed by multiplying the value per minute of the constraint by the total number of minutes taken by the hole. The value per

Constraint Schedule

Figure 11   Holes in the schedule result in lower throughput.

$$\frac{\text{Total Sales}}{\text{CCR Time}} = \frac{\$300,000}{6000} = \$50/\text{min}$$

$$\text{Unused CCR Time} = 30\text{min}$$

Estimated value of unused CCR time

$$30 \text{ X } \$50 = \$1500$$

Figure 12 Estimating the value of unused CCR time.

minute of the constraint is the total sales generated by the schedule divided by the total time of the constraint used to generate the sales (see Fig. 12).

If the schedule for the constraint can be successfully rebuilt so that the hole no longer exists, then the issue in determining the impact of an off-load to a vendor is comparing the needs of the customer (having an on-time delivery) to the additional expense created by sending the order out. If there is an intolerable number of past due orders created on a consistent basis, then thought should be given to either elevating the constraint or increasing the amount of protective capacity on the secondary constraint so that conflicts are eliminated.

Note: After the constraint is rescheduled a second attempt is made to subordinate the remaining resources. If the sales orders involved in the conflict are pushed into the future by a full rod or buffer length, the same conflicts should not recur.

C. Off-Loading from Resources that Use Excess Capacity

Whenever excess capacity is present it should be recognized as unrealized throughput. In other words, there is value in the amount of throughput that could be generated by the excess capacity. If excess capacity is used to generate sales, the value of the order is considered the difference between the sales price and the cost of raw material. This can be a tremendous competitive edge and if it's not being used, someone should be attempting to identify how it could be.

However, to understand the impact of excess capacity on the purchasing decision some key questions should be answered:

- What is the impact of receiving material late on resources that have excess capacity?
- What is the impact of paying more for the part?

- Will receiving the part late create a temporary bottleneck? If so, what is the impact on those orders that go through resources which have the temporary bottleneck and are later processed on the constraint?

Normally, for those products that are processed entirely on resources where excess capacity is available and the part is delivered late, there is enough capacity to process the order and still deliver on time. If not, there will be enough capacity available at a later time to make up the difference in the total amount of throughput generated. The value of the raw material part to the company is less when excess capacity is involved. Since throughput is defined as the sales price minus the cost of raw material, paying more to obtain a part that is processed with excess capacity will result in an immediate decline in throughput.

Whenever raw material parts arrive late the effect is the same as shrinking the buffer. When the buffer is shrunk and parts are released later, the resource load is increased by having to accomplish the task in less time. This situation can create temporary bottlenecks threatening the constraint's ability to deliver on time. However, the normal buffer management system is designed to handle most late deliveries. For those parts that are processed on the constraint and go through resources that have a temporary bottleneck, priority should be given to those parts that create holes in the buffer (see Fig. 13). Resource R-3 is shared by part/operations that feed sales orders S/O111 and S/O222. Only sales order S/O111 is fed by the constraint. Raw material part A is used exclusively on S/O111 and raw material part B is used only on S/O222.

Due to a problem in delivery from the vendor, Part A was released late for processing on R-4 and now, since B arrived first, A must wait until B is finished processing. This is a temporary condition. In other words, the total available capacity on resource R-4 is more than enough to fill the needs of

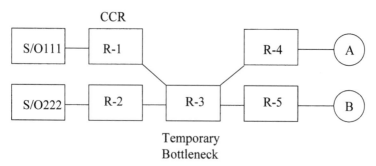

Figure 13 Purchasing decision with bottleneck in schedule.

protecting the constraint. However, because A was late R-4 is temporarily required to exceed its capability. Since this is physically impossible, a delay in the arrival of A to the constraint is imminent. If the delay is long enough that it begins to threaten zone I of the buffer at R-1, then part A will be expedited, thereby alleviating the situation. In this case paying more for the part that feeds the nonconstraint resources, which uses excess capacity so that parts will arrive on time up to a point, is a waste. At what point should the additional expense be approved?

The buffer management system discussed in Chapter 9 is used to manage the situation where normal fluctuations in the production process occur during execution of the schedule. However, by looking at the release schedule for raw material a decision can be made earlier in the process, when it is known that there is a high probability that a specific delivery date will cause a problem.

Figure 14 is a raw material requirements schedule. Included in the schedule are

- Part number and description of the part required
- Gating operation and resource where the part is to be released
- Buffer size for the buffer origin that the specific part/operation feeds
- Scheduled release date at the gating operation
- Due date of the purchase order required to fill the requirement
- Percentage of the buffer taken

Notice that this report is in descending sequence based on the amount of buffer taken by the late delivery. Raw material part 12345 for gating operation 123/30 on resource R-7 is scheduled to be released at day 154 and feeds a 5-day

Late Raw Material Requirements Schedule

Part No.	Description	Gating Op./ Resource	Buffer/ Size	Release Date	Due Date	% Buffer
12345	Armature	123/30/R-7	C/5 Days	154	158	80%
12369	Gasket	124/30/R-7	C/5 Days	154	157	60%
45783	Flight Line	234/40/R-6	A/5 Days	153	155	40%
27240	Sky Hook	456/30/R-8	S/5 Days	152	153	20%

Figure 14 Late raw material requirements schedule.

buffer for the constraint. However, it is not scheduled to arrive from the vendor until day 158 and has taken 80% of the buffer in front of the constraint. In this case the part must be expedited. Part 27240 is going to be late, but it takes only 20% of the shipping buffer. In this case the part does not justify spending additional money to expedite.

 Note: Under the "Buffer/size" column, the constraint, shipping, and assembly buffers are designated C, S, and A, respectively. For example, "A/5 Days" means a 5-day assembly buffer.

IV. SALES AND MARKETING

The TOC-based information system can have a very large impact to the sales and marketing department. When the impact of the TOC approach is compared to the traditional method the difference is quite noticeable. Poor scheduling methods and cost accounting have caused major problems for these two organizations. The information system will be discussed with the following tasks in mind:

- Selling the right product mix
- Accepting orders
- Quoting delivery dates
- Changes in the schedule
- Selling excess capacity
- Developing new markets

A. Selling the Right Product Mix

Selling the right product mix can have a tremendous impact on profitability, yet most companies do not have the slightest idea of how to perform this task. This is not surprising considering most people have been taught incorrectly for the last 50 or more years. In addition, there are very few information systems being sold in the world today that can help to perform this task. Without rehashing the decision process, what is needed to make decisions is a comparison of the throughput generated per unit of the constraint for the products being sold.

 What the information system must provide to determine the right product mix to sell into the market is

- The identity of the constraint
- The amount of time each product absorbs from the constraint
- The sales price of each product
- The raw material cost

Most systems today can provide the last two ingredients. Because of the problems discussed in Chapter 2, few systems can provide the first two. Once the constraint has been identified, what is needed is a method for calculating the total time for all part/operations that cross the constraint and belong to a specific product. It is a simple task to create a list of products with their respective throughput per unit of the constraint figures in descending order, the top being the most profitable and the bottom the least (see Fig. 15). Product A is the most profitable, with $10.00 per minute, followed by B with $8.00. Product D is the least profitable. When setting sales strategies and commission schedules this information is vital and should be readily available from the information system.

1. When Labor Is the Constraint

Companies will often assume that a constraint is a physical resource or machine in the factory and that it is constantly moving from resource to resource depending on product mix. To adapt, employees are trained to handle numerous jobs so that when the constraint moves the workforce can be adjusted to fit the new environment. When the number of employees is no longer able to meet the demand by switching from resource to resource, the available labor generated by employees becomes the constraint. This circumstance is rarely observed correctly. Companies will usually continue in the attempt to move employees as the constraint moves, not realizing that the constraint is actually the number of production people available.

Determining product mix in this environment is very similar to the way in which it is calculated when the physical constraint is a machine. The amount of time used by the constraint is divided into the amount of

Product	Sales Price	Raw Material	Throughput Generated	Total CCR Time	T/uc
A	300	150	150	15	$10.00
B	310	150	160	20	$8.00
C	200	100	100	15	$6.67
D	400	150	250	45	$5.56

Figure 15 Determining the right product mix.

Product	Sales Price	Raw Material	Throughput Generated	Total Labor	T/uc
A	300	150	150	60	$2.50
B	310	150	160	120	$1.33
C	200	100	100	30	$3.33
D	400	150	250	90	$2.78

Figure 16 Product mix data showing the total labor required per product.

throughput generated. The difference is that the constraint is the total employee time required to produce the product. Figure 16 is similar to Fig. 15 in that the products, sales prices, raw material costs, and throughput are the same. However, notice that the time used for the constraint has increased to include the total labor required to produce the product. Notice also that the order of the most profitable product has changed. Product C is now the most profitable, while product B is the least. Whenever the constraint changes there will probably be a change in product mix.

  Note: It has never been the constraint's obligation to tell anyone that it has moved.

2. When Raw Material Is the Constraint

Sometimes the constraint will surface in the form of raw material. In other words, there will not be enough of a specific part to build all of the requirements of the market. In Fig. 17, raw material part A is contained in products F, I, H, and G. There is a limit to the number of As available to fill the market for these products. To compute the throughput per unit of the constraint, the number of parts required for each product is divided into the throughput generated for the product. In this case, product F is the most profitable. It generates $300 in throughput each time one is sold and it takes three part As to create each one. The throughput per unit of the constraint is therefore $100.

B. Accepting Orders

The key factor in accepting a sales order is identifying the lowest amount of throughput per unit of the constraint that is being sold now. As an example, if

Raw Material Constraint: [A]

Product	Throughput	Req/Unit	T/uc
F	$300	3	$100
I	$100	1	$100
H	$150	2	$75
G	$200	3	$66

Figure 17   Product mix when raw material is the constraint.

the constraint is being sold for $6.00 per minute now (by definition it has no additional capacity), to accept an order at less than $6.00 a minute would decrease the profit of the company. In Fig. 15, if all the constraint's time were spent processing products A, B, and C, processing D would cause profit to immediately decline.

To make this decision the salesman needs to know

- The amount of throughput per unit that would be generated from the product being sold
- The lowest throughput per unit of the constraint that is being sold now

Salesmen should go into the field knowing these two facts, and management should ensure that they are provided.

C. Quoting Delivery Dates

When quoting a delivery date for products that go through the constraint the salesman only needs to look at the constraint schedule for open time on the horizon in which to fit the order. For a precise schedule, the system should allow the processing of sales quotes that could load the system with data as if it were a valid order and have it process the new delivery date based on the additional data. (This requires a different system structure, which will be discussed in Chapter 12.)

If the salesman is selling excess capacity, it is important not to create temporary bottlenecks on the nonconstraint resources that will threaten the constraint. It would still be necessary to load the quoted order on the system to arrive at a precise delivery date.

D. Changes in the Schedule

There are two schedule-related conditions under which a change in order due date will occur:

- Those changes which happen in the schedule and are predicted beforehand
- Those that occur as a result of problems which create holes in the buffer

When changes occur in the schedule due to conflicts between resources or a lack of adequate protection, this information is immediately fed back to the master schedule. Because of this feedback mechanism, a salesman can be made aware of these conditions immediately and is able to make priority decisions on which orders should actually slip (see Fig. 18). Order 2 (connected to sales order S/O111) and order 4 (connected to sales order S/O222) are on the constraint schedule. Order 4 is going to be late because it has taken 75% of its buffer time. Order 2 is actually early; it has an additional 10% of the buffer length in front of it. If rescheduled to a later time, sales order S/O222 will probably be pushed so that the full buffer in front of order 4 is recovered. If the length of the buffer is estimated to be 2 days, the customer can be notified that the order will be 1½ days late.

However, what if the customer for sales order S/O222 could not really afford to have the order postponed and the customer for sales order S/O111 could? By giving order 4 priority over order 2, 150 minutes of constraint time can be recovered and given to order 4 along with part of order 2's buffer

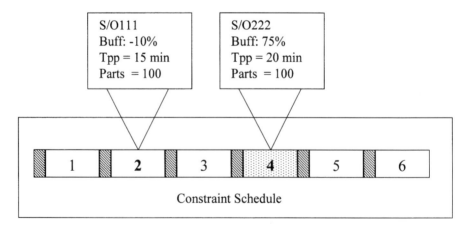

Figure 18   Information of schedule changes helps decision-making.

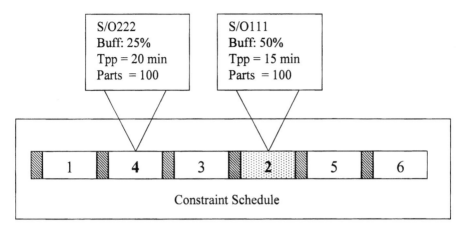

Figure 19 Quoting delivery dates.

(see Fig. 19). The buffer for order 4 was partially restored while the buffer for order 2 is now at 50%.

As discussed earlier, orders that cause holes in the buffer for zone I will be given a higher priority over those that are on time. This priority is given by having the offending order expedited. If during the expedite phase it is discovered that the order will definitely not meet the shipping schedule, then the sales department should be notified immediately and feedback given to the customer.

Note: It is better to tell the customer early that an order will be late rather than waiting until it is already late.

Setting the new ship date depends on where in the process the order is and how far it must travel before getting to shipping. If the order is being expedited, it will take less time than the normal buffer length to reach the buffer origin. If the buffer is for the constraint, there may be additional capacity afterward so that the order may still reach shipping on time.

When a problem occurs on R-5 in Fig. 20, causing a hole in the buffer for R-3, and it is actually late leaving the constraint, there is also the buffer in front of shipping to consider.

The maximum that an order should take is one shipping buffer after leaving the constraint. The actual due date given the customer is based on the experience of the buffer manager in judging how long it will take to process from a given operation.

Once an initial due date is determined it can be validated by placing the new due date into the system and creating a new schedule.

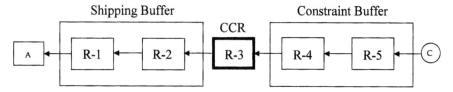

Figure 20   Impact on the shipping buffer.

Note: In creating the alternative schedule, the original schedule should not be changed. Provisions should be made to copy the original schedule and modify it to determine if the new schedule is reliable. By changing the old schedule, the order would not appear as late and, therefore, would not be expedited (see Chapter 12).

E.   Selling Excess Capacity

For sales, the absence of an internal constraint is usually an indication that a policy constraint is blocking them from pulling the physical constraint inside the company. However, the presence of a physical constraint does not indicate that all excess capacity has been located and a plan developed to exploit it. Before excess capacity can be sold, management must know

- Where the excess capacity is
- How much of it is available
- What products are impacted

Excess capacity cannot be found based on the amount of idle time recorded at any given resource. Since the scheduling process involves all resources within the throughput chain, what appears as excess capacity at one resource may actually be capacity used to protect another. In Fig. 21, resource R-1 is the constraint and is loaded to 100%. It will obviously have difficulty delivering to the sales order more often than a resource that is loaded to 80%.

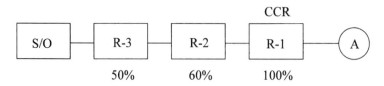

Figure 21   Selling excess capacity.

However, it is not just resource R-1's effect which is involved. If the order is late leaving the constraint, resources R-2 and R-3 have additional capacity to devote to alleviating the situation. In a chain of resources one resource cannot be judged individually. Their performance must be judged as a team.

Note: A throughput chain involves all resources that are connected from the vendor to the market by the process of generating throughput.

If a string of resources has the capability to deliver to the market more than the market is willing to buy, then there must be excess capacity. This is a key issue. The question is how much additional capacity is there and how should it be used? When the constraint is in the market, the system will attempt to subordinate to the schedule created by the sales orders. When the system can no longer accomplish this task without increasing capacity on an internal resource (exploitation), then the company is no longer selling excess capacity for that throughput chain.

F. Developing New Markets

The TOC-based system can simulate reality so well that capacity simulations can easily spot where additional resource availability can be found for certain products and where others will create conflicts. If in the process of determining how to break a certain marketing constraint new products are proposed, the impact of the new products and how they will affect the internal capabilities of the manufacturing resources can be predicted. Mock bills of material, routings, and a forecast can be used to simulate the new load even before the products are designed.

Note: If one throughput chain has a capacity-constrained resource, it does not mean all throughput chains within the company have one. If there is no excess capacity in one area of the company, look in another.

V. FINANCE AND ACCOUNTING

A. Throughput Justification

One of the biggest advantages of using the TOC-based system for finance and accounting is the ability to really examine requests for capital expenditures. Since cost accounting has major flaws in this area (e.g., cost justification), the need for this kind of tool in justifying spending is quite large.

In cost justification the objective is to justify an expense based on the amount of cost that is reduced. Seldom is the actual cost reduction ever realized. As an example, whenever a justification is made for the purposes of buying the traditional information system it is usually based, at least partly, on the increase in the efficiency of employees. This increase in

Cost Justification for an MRP System

Inventory Value	$10,000,000		Savings	
Decreased Inventory	20%	Inventory	$160,000/YR	
Carrying Cost	8%	Labor	$65,000/YR	
Cost Reduction	$160,000/YR			
Labor Cost/Inventory Management	$150,000/YR	**Total Savings**	**$225,000/YR**	
Increased Efficiency	10%	**Investment**	**$500,000**	
Cost Reduction	$15,000/Yr			
		Break Even	**2.2 YRs**	
Labor Cost/Production	$1,000,000/YR			
Increased Efficiency	5%			
Cost Reduction	$50,000/YR			

Figure 22   Cost justification of an ERP system based on labor savings.

efficiency is converted to labor savings and presented as a reduction in cost (see Fig. 22).

Unfortunately, this labor savings is almost never realized. Instead, the company pays for the system, which increases the money leaving the company and never has a layoff to offset for the increase in expenses. Additionally, because of the problems discussed earlier, MRP can have a damaging impact on the scheduling of the factory. If an increase in rate at which inventory is consumed does not occur, then inventory may not decline. In fact, it may even go up.

In order to justify spending money the justification must be based on the increase in throughput generated by the expense. Justifications can fall into three categories:

1. Those used to exploit or elevate the constraint
2. Those used to increase protective capacity
3. Those used to increase excess capacity

The first category will obviously result in throughput being increased. When the amount of increase in throughput due to the change is compared to the cost of the change, a real payback figure can be

Additional
Throughput $10,000/mo.
Generated

Cost of
Improvement $60,000

Payback   6 mo.

Figure 23 Throughput justification.

obtained (see Fig. 23). The additional throughput generated is estimated to be $10,000 per month due to the improvement. The cost of the improvement is $60,000 and the payback period is 6 months. This is the minimum information needed by the finance department when justifying an expenditure and should be requested from all departments when submitting.

1. Creating the Throughput Justification (Improving Without Breaking the Constraint)

In creating the throughput justification for improvements to an internal capacity-constrained resource the key issues to understand are

- How much additional capacity will be generated on the constraint by making the improvement
- What will be done with the additional capacity

By definition, unless the constraint is broken, there will always be more market available than productive capability. However, if sales is doing its job, those products which create the highest throughput per unit of the constraint will be limited by the market and not the capacity-constrained resource.

If multiple products are being scheduled on the constraint and one is more profitable than the others, marketing's objective should be to sell all the available constraint time for the most profitable product until it has reached the marketing constraint for that product. Once reached, the additional capacity should be taken by products that are of less value. So in trying to determine how much additional throughput will be generated by an improvement, the additional throughput will, more than likely, be originating from those products that are less profitable.

In Fig. 24, R-4 is the capacity constrained resource. Resource R-4 processes products L through O and is loaded with 4755 minutes for a 10–7 week period. Product M is the most profitable at $10 per minute of constraint time, while product O is the least profitable at $3.33 per minute. Notice that after an improvement has been made in R-4's capability the processing time for product M has dropped from 20 minutes to 10 minutes per unit under the new time column. This means that the 120 units of product M now take 1200 minutes to make instead of the original 2400. The question is what should be done with the additional 1200 minutes released from the constraint?

All of the available market for products M, N, and L should have been already sold or an order would not have been accepted for product O. Since the basis of the example was to improve the constraint and not to break it, there should be more market for product O than available capacity. In justifying an expenditure for increasing capacity on resource R-4, all 1200 minutes of additional capacity created by reducing the amount of time taken to produce M would be absorbed by selling additional Os. Throughput per minute of the constraint for selling and producing product O is $3.33. Multiplied by the additional capacity released (1200 minutes), the additional throughput generated is $3996 for the 10-week period. If the investment is $10,000, it would take approximately 6 months to break even.

If a marketing constraint is encountered prior to reaching the end of the additional 1200 minutes, then the physical constraint for R-4 has been broken and a new constraint must be identified and included in the justification of the $10,000 investment. Since the constraint has been broken, a new product mix must be established before any throughput estimate can be made. However, it may be that part of the justification will

Resource R-4

Product	T/uc	Time/unit	Qty	Total Time	New Time/Unit
M	$10.00	20 min	120	2400	10
N	$6.66	45 min	37	1665	45
L	$5.55	90 min	5	450	90
O	$3.33	30 min	8	240	30

Figure 24   Throughput justification—redistributing processing time.

include the introduction of a new product that will absorb the remaining constraint time and leave R-4 as the primary constraint.

2. Breaking the Constraint

In breaking the constraint the change in the company can become quite complex and have major implications. Key issues in determining the impact include the following:

- If the constraint is to be broken, where will it go next and what will be the impact on the current load in the factory?
- How much will throughput go up based on the current demand?
- What list of products will now become the most profitable based on throughput per unit of the constraint?
- Will any new conflicts develop between other resources?
- Can an effective schedule be created given the change?
- Will the constraint be pushed into another department or possibly into the market? If so, what will be the impact?

To answer these kinds of questions will take an information system with the ability to simulate the change. Fortunately, while the change may be complex, the process to be performed never changes. The objective should be to implement the five steps given the new data.

The first issue is to make the change in the data used by the system and see where the new constraint might go. Once identified, the remainder of the five steps must be implemented. After completion, the impact of the change can be determined and the justification either approved or disapproved.

3. Increasing Protective Capacity

If the request is to facilitate subordination by increasing the amount of protective capacity available, one of two situations should exist:

1. A persistent conflict between two resources which has gone unresolved and results in continuously rescheduling sales orders to later time periods
2. Holes in the buffer which can be directly attributed to a specific problem

In solving conflicts that result in orders being pushed into later time periods, the justification is made based on the value of the sales dollars pushed. For those situations where a specific problem results in constantly expediting material due to holes developing in zone I of the buffer, the justification is made based on the amount of inventory or operating expense reduced as a result of being able to shrink the buffer due to the change.

4. Increasing Excess Capacity

The third situation is usually presented by the person who does not have experience or knowledge of TOC. The justification is usually based on the amount of time saved by not doing a specific activity. While this can usually be dismissed quickly, it is an indication that someone in the organization needs some training. Unless there is a valid plan to use the excess capacity to increase throughput the request should be turned down.

B. Cash Flow Management

Having a valid schedule with predictable results means being able to effectively manage cash flow. This is extremely important for companies that are having problems in this area. The system can be used to predict the rate at which money enters and leaves the company so that better projections can be made. By simply assigning a throughput value to each sales order, the schedule will produce a forecast of the rate in which money enters the company. Notice that if labor and overhead are removed from the current cost accounting system, the cost of goods sold is now the cost of raw material. Subtract that figure from the sales price and the result is the throughput figure for the company. Notice that this figure will also be the profit margin on most traditional systems.

A raw material receipt schedule along with the current operating expense plus the overtime projected by the system will identify the cash needed to operate the company. Operating expense can be adjusted based on the additional overtime required by the schedule.

The result is a complete picture of how and when money enters and leaves the company for the planning horizon.

VI. QUALITY ASSURANCE

The constraint should be fully utilized in generating throughput. From a quality perspective this means

- Parts processed by the constraint should not be delayed by rework.
- After being processed on the constraint parts should not be scrapped.

Minimizing inventory and operating expense means reducing the problems that prevent the reduction of protective capacity (those things that create holes in zone I of the buffer).

In addition to identifying the constraint and providing the buffer management process, the system should supply an idea of how resources interface. In other words, which part/operations and resources

- Feed the constraint
- Are diverging
- Are converging
- Appear after the constraint

Trying to produce product flow diagrams to define these issues can be a tedious effort and will probably require revisions as soon as the ink is dry on the first set. Two visual tools supplied by the information system can help tremendously in this effort. They include the "where used" and "descending part/operations" screens.

A. The Where Used and Descending Part/Operations Screens

These screens are actually visual representations of the net. The descending part/operations screen starts at the sales order and presents all part/operations and the resources on which a specific sales order is to be processed. The where used screen starts at the raw material and traces upward through all part/operations and resources to all sales orders which may be using the part (see Fig. 25).

Notice that the constraint appears in bold to readily identify if there is a problem that exists on a resource prior to or after the constraint. For example, Fig. 25 shows a capability index (Cpk) of 1.00 on resource R-5

Descending		Where used	
S/O111 - A		RM C	
A/30	R-1	B/10	R-6
A/20	R-2	B/20	R-5
A/10	R-3	B/30	**R-4**
B/30	**R-4**	A/10	R-3
B/20	R-5	A/20	R-2
B/10	R-6	A/30	R-1
RM C		S/O111 - A	

Figure 25 Descending part/operations and where used screens.

Red Lane   Feeding the Constraint

S/O111 - A	
A/30	R-1
A/20	R-2
A/10	R-3
B/30	R-4

B/30	R-4
B/20	R-5
B/10	R-6
RM C	

Figure 26   Identifying the red lane and those resources feeding the constraint.

part/operation B/20; it is easy to see that B/20 feeds the constraint and that a low Cpk could cause some serious problems for the constraint's ability to perform. It is also easy to see that if any scrap occurs on part/operations B/30 through A/30, the loss would be the price of the entire sales order. In addition, if there is a problem in receiving a particular purchase order on time or if there is a quality problem which affects raw material part C, the where used screen can identify the part/operations and sales orders that are impacted.

An abbreviated version of these screens includes those part/operations and resources that appear after the constraint and feed the sales order (the red lane) as well as those part/operations and resources that feed the constraint (see Fig. 26).

A modification of the descending part/operation and where used screens includes the addition of some important indicators (see Fig. 27). Notice that

PRODUCT A   PRODUCT B
PRODUCT C

Resource	Scrap	Rework	Down Time
R-1	.01		
S R-2	.05	.10	.07
R-3	.00		
P R-4	.02	.03	.05
R-5			
R-6			

Figure 27   Key indicators in the throughput chain 1.

Throughput Chain 1

Part Op	Qty	Rwk	Scp		PRODUCT A		PRODUCT B	
I /30	4000	150	100		PRODUCT C			
G/30	3500	0	0		Resource	Scrap	Rework	Down Time
C/30	3000	0	0		R-1	.01		
A/30	2500	375	250		S R-2	.05	.10	.07
B/30	2000	0	0		R-3	.00		
H/20	1500	0	0		R-4	.02	.03	.05
J /10	1000	0	0		R-5			
					R-6			

Figure 28   Inquiring on the details.

rework and downtime information only appear for resources R-2 and R-4. To place this information on the system would only cause confusion if there is capacity available to handle these issues. Additionally, if there were a problem on these resources, it would appear in buffer management. The objective of this screen is to point the activity toward those resources that have critical capacity shortages. Scrap appears only for those resources that go from the constraint to the end of the process (the red lane).

These are important keys in focusing the quality program. But additional information is still needed, such as on what part/operations the scrap or rework is occurring. In Fig. 28, part/operations I/30 and A/30 are absorbing all of the scrap or rework time.

B. Making the Scrap or Rework Decision: Revisiting the Quality Decision

A decision that can involve both the production and the quality organization is how to know when to scrap or rework a part. Using the traditional methods will not work. Figure 29 shows the data required to make the decision and can serve as a basis for what features might be required for the information system to help in this process.

As previously indicated, product A takes 15 minutes to process on the CCR and uses $25 in raw material for each part produced. An order of 100 is

PRODUCT	CCR TIME/UNIT	CCR TIME AVAIL
A	15	1500 MIN
ORDER = 100	SP = $100	RM. = $25

DEFECTIVE PARTS	TIME/UNIT	MATERIAL
10	10	$5

Figure 29 Scrap versus rework revisited.

being processed at a sales price of $100 each. After 50% of the order has been processed, ten parts are found to be defective. It will take an additional $5 in raw material and 10 minutes of constraint time to repair each one. The question is should the ten parts be scrapped or reworked?

If the ten parts are to be scrapped, then they will need to be replaced to fill the order. Sixty additional parts are to be processed creating throughput of (100 − 25) $75 each for a total throughput generated of $4500. Since each part takes 15 minutes of constraint time and there are 60 parts, the total constraint time consumed will be 900 minutes. To replace the raw material scrapped for each of the ten parts, $250 must be subtracted from the total throughput generated. To scrap and replace the ten parts and process the remaining order will take 900 minutes and generate $4250, or $4.72 per minute of constraint time (see Fig. 30).

To rework the ten parts requires an additional $5 raw material be subtracted from the sales price (100−30), which will generate $70 of throughput for each of the ten parts processed for a total throughput of $700. It will take 100 minutes of constraint time to process all ten parts. The remaining 50 parts will take 750 minutes of constraint time and generate $3750, for a total of $4450 generated with 850 minutes of constraint time, or $5.23 per minute (see Fig. 31). The throughput per minute of the constraint for the rework is greater, so the choice from the data given would be to rework the parts.

The decision can be made with much less data than given. Once the parts causing the problem have been processed on the constraint, the question

				(T)	CCR TIME
THROUGHPUT	=	60 (100 - 25)	=	$4,500	900
MATERIAL	=	10 X 25	=	$ 250	
				$4,250	900

$$T/uc = \frac{\$4,250}{900} = \$4.72$$

Figure 30 Scrap versus rework revisited. Throughput per unit of the constraint if parts are scrapped.

becomes how to maximize the throughput of the system given the current situation. The first issue is that the parts must be replaced so that the sales order can be delivered. At first glance it will take $5.00 of additional raw material and 10 minutes of the constraint's time to generate $100 for each part if the parts are reworked. If they are scrapped, it will take an additional $25 of raw material and 15 minutes of constraint time to make the $100. Throughput per unit of the constraint for rework is $9.50. Throughput per unit of the

				(T)	CCR TIME
THROUGHPUT	=	10 (100 - 30)	=	$700	100
THROUGHPUT	=	50 (100 - 25)	=	$3,750	750
				$4,450	850

$$T/uc = \frac{\$4,450}{850} = \$5.23$$

Figure 31 Scrap versus rework revisited. Throughput per unit of the constraint if parts are reworked.

				(T)	CCR TIME
THROUGHPUT	=	10 (100 - 5)	=	$950	100
THROUGHPUT	=	10 (100 - 25)	=	$750	150

Rework   Scrap

$$T/uc = \frac{\$950}{100} = \$9.50 \qquad T/uc = \frac{\$750}{150} = \$5.00$$

Figure 32 Scrap versus rework revisted—calculations for scrap versus rework.

constraint for scrap is $5.00 (see Fig. 32). The only data necessary to make the decision are:

- The original cost of raw material for the product
- The sales price of the order
- An estimate of the amount of additional raw material and constraint time required to repair
- The quantity of the parts to be reworked or scrapped
- The original constraint time for the part

The question now is what can the system do to provide this kind of information? Figure 33 represents an interactive screen that can be created

			T/uc
Part/Operation	A/30	Scrap:	$5.00
		Rework:	$9.50
CCR Time Est. (Rework)	10		
		Product:	A
Add. Raw material (Rework)	$5.00	Sales Order:	S/O111

Figure 33 Scrap versus rework—providing information to make decisions.

to help with the scrap/rework decision. The only data needed to be entered are the part/operations being considered, the estimated constraint time to repair, and the additional raw material required. Additional data, such as the sales price of the order, the original constraint time, and the original raw material cost, can be obtained from the manufacturing system database.

1. The Impact of Scrap and Rework on the Scheduling Process

Whenever a scrap or rework occurs there will be an impact on the amount of labor and material required by the system. This must be reflected in the scheduling process. Figure 34 is a product flow diagram for product A, which has a sales order demand for 10 units. The numbers appearing between each part/operation are the bill of material relationships. Notice that for A/30 to be processed it must receive one A/20 from resource R-2. Notice also that resource R-2 scraps 25% of the A/20s it creates. This means that for every A/30 created, 1.33 A/20s must be processed first. To process 1.33 units of A/20, R-2 must receive 1.33 units of A/10. To produce all ten units of A on the sales order requires 13.3 units of raw material C.

Scrap affects not only material requirements, but resource requirements as well. In order to obtain one unit of A/20 to be transferred to A/30, resources R-2 and R-3 must process the additional material. Additional processing will obviously take more resource time. This kind of processing must be considered when creating the net and in determining the load. It is done by modifying the bill of material relationship during processing.

Rework is considered somewhat differently than scrap. Scrap is simply yielded during processing. However, rework will usually require a modification of the net (see Fig. 35). Notice that 25% of all A/10s are reworked back to the beginning of the process. This means a certain increase in the load for resources R-3, R-4, and R-5. It will also mean an adaptation of the rods, provided a constraint schedule is involved, to ensure that reworked material maintains a specific distance in time from the process time of the original parts.

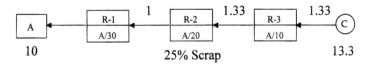

Figure 34 The impact of scrap on the scheduling process.

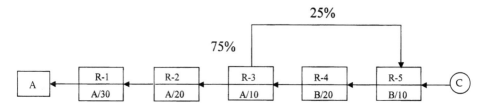

Figure 35   The impact of rework on the scheduling process.

VII. ENGINEERING

As described in *The Theory of Constraints: Applications in Quality and Manufacturing*, the focusing mechanisms provided by the traditional methods, such as single minute exchange of die, total productive maintenance, statistical process control, design of experiments, and quality function deployment, lack effective mechanisms for focusing activity so that the use of the tool is in line with the goal of the company. The TOC-based information system can provide that focus. From a physical perspective, the task of determining what to fix and when to fix it is made easy.

     To increase throughput these tools should focus on the constraint(s). To decrease inventory and operating expense they should be focused on those things that consistently create holes in zone I of the buffer. However, it is not enough just to know that a reduction in setup time will increase throughput. The question that should be asked is how can engineering minimize the amount of time spent in setup or machine downtime in the generation of throughput.

A. Setup Reduction

Figure 36 is a Pareto analysis of the setup time for those part/operations that cross the constraint. By pinpointing those part/operations that take the most time from the schedule to perform setup, engineering's time can be better focused to maximize the generation of throughput.

     Resource R-4 is the primary constraint, with 7000 minutes total demand. After having completed the schedule generation to include setup reduction, the amount of run time available to generate throughput is 86% of the total time available. This means the constraint is losing 14% of its productive capability to setup. Notice that 30% of the 1000 minutes is being used for part/operation E/30. By concentrating on reducing the setup requirements on part/operation E/30 and making significant reductions, engineering will return more throughput for its effort.

RESOURCE R-4

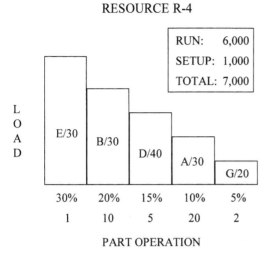

Figure 36   Setup reduction.

An additional consideration is the volume of setups performed. If E/30 takes 30% of all setup time on the schedule but only one order is involved, the opportunity for improvement may be less than for part/operation B/30, which includes ten setups. Notice that under each percentage in the figure the number of setups involved is also shown.

Note: A high number of setups also indicates that setup savings may be extended to cover more orders.

B.   Total Productive Maintenance

As seen in *The Theory of Constraints: Applications in Quality and Manufacturing*, total productive maintenance (TPM) has some serious problems in attempting to focus on the right areas of the company. Like most cost based focusing mechanisms, TPM uses local measurements without considering that the resources involved are somehow connected to one another. The TPM effectiveness rating is comprised of the measurements of availability, efficiency, and quality:

Effectiveness = availability × efficiency × quality

Each resource is measured in this fashion regardless of where it is in the overall throughput chain or what the characteristics of the individual resource might be.

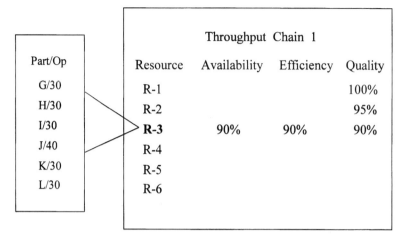

Figure 37 Refocusing total productive maintenance.

To be effective, the individual measurements of availability, efficiency, and quality must be placed in the context of the way in which resources interface. In other words, what are the relationships between resources and how do the individual measurements help in determining where to focus specific efforts to improve?

In Fig. 37, notice that resource R-3 is in bold and that availability, efficiency, and quality ratings are shown. Resource R-3 has been identified as the constraint. Notice also that the chain of resources leading from R-3 to the finished product have quality ratings. The objective for this screen is to identify for the engineer where to place efforts to maximize the amount of throughput generated. To do this, availability, efficiency, and quality should be maximized on the constraint. Quality must be maximized to prevent scrap on the string of resources leading to the finished product. Additional information provided in the window to the left includes the part/operations that cross the constraint.

Note: To improve on the remaining resources, buffer management is used so that the impact of the problems encountered on the nonconstraints can be aggregated.

C. Product Design

Whenever efforts are necessary to improve the amount of throughput generated by the constraint, a key issue in releasing the maximal amount of

Resource R-4

Part/Op	PROD	T/uc	Time/ Unit	Units Produced
I /30	I	10.00	0:30	4000
G/30	G	9.00	0:45	3500
C/30	C	15.00	0:20	3000
A/30	A	8.50	0:25	2500
B/30	B	14.00	0:30	2000
H/20	H	6.00	0:45	1500
J /10	J	7.00	0:45	1000

Figure 38 Support for engineering.

constraint time for the effort given is to design products so that they actually take less constraint time. Figure 38 shows a list of all part/operations that cross the constraint, the product produced, the throughput per unit, the amount of time it takes to produce each product, and the number of units to be produced on the schedule.

Notice that the chart is listed in descending order based on the total number of parts to be produced on the constraint. For each minute saved in producing product I, 4000 minutes of constraint time will be released to sell into the market. The question then becomes what products should be sold given the additional 4000 minutes. This, of course, is a marketing issue.

VIII. PRODUCTION

Like every department within the company, the decisions made by production should be based on those actions required to implement the five-step process of improvement. Decisions should obviously be made in the context of where in the process production is. If the constraint is in the market, then the actions required will be different than if the constraint is on a particular resource within the plant. The first step, regardless of where it is currently located, is to identify the constraint and then subordinate the actions of the other departments.

With one exception, most of the decision support for production has been applied in the creation of the schedule and control portions of the system. This includes

- Setup savings
- Order priority
- Overtime
- Expediting
- Head count requirements
- Off-loading

The system prompts the user at certain times to make decisions regarding the task being performed. As an example, when the constraint has been identified and the first step in the exploitation process has been accomplished, a schedule exists which includes a list of orders with start and stop dates for specific part/operations across the constraint. What also exists is an indication of what orders will be late unless changes are made. The decision of how to fix the late orders may include setup savings, off-loading of orders, or overtime.

During the scheduling process the system offers an opportunity to perform any of these tasks and will demonstrate the outcome through simulating the interactions between the resources as a result of the decision. However, it does not indicate which should be done first or in what order. This is a management decision.

The preferred method of presentation for the schedule as well as a model for the decision process is graphical. The system will present orders from the schedule on a timeline for the constraint being exploited. It will show which orders are on time and which will be late based on color. In Fig. 39 red indicates late and green indicates an order that is on time. In the figure, orders 1 and 4 are indicating that they will be late. The determination is made by the amount of buffer still remaining in front of the order. If an order has less than half the buffer remaining, it is considered a

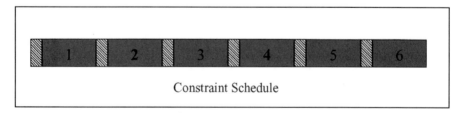

Constraint Schedule

Figure 39   Schedule presentation.

problem order and will be late. A decision must be made to determine the best way to exploit the resource.

When making a decision that affects the exploitation process, a certain pecking order is indicated by the impact to profitability. Setup savings should be tried first. It is the cheapest way to increase capacity. Next comes off-loading, then overtime, and, as a last resort, postponing certain sales orders.

However, each decision cannot be made in a vacuum and must be tempered by the situation. As an example, the system has no way of knowing that part of the data it may be using is wrong. If an indication is made by the system that a conflict exists between the primary and secondary constraint and off-loading should be considered, the system may not know that the particular resource needed for off-loading is broken and that the alternative decision to push sales order due dates may be a better solution. Additionally, the system can indicate which orders are going to be late as a result of the conflict and is able to relate the new due date of the sales order, but it can't understand which customer orders should be postponed at the expense of others. This is why human intervention is necessary during certain steps in the scheduling process.

A. Off-Loading

To aid in the off-loading of orders the system should provide a method of determining which alternative resources are available so that when making the off-load decision, possible selections will be readily available (see Fig. 40).

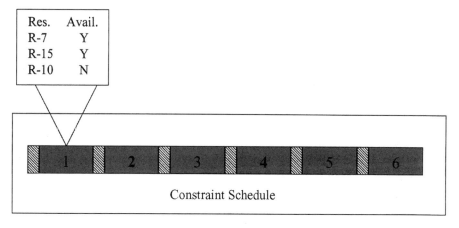

Figure 40 Off-loading constraint schedule.

What the system can offer is alternatives. And because it resembles reality so well, it can simulate the impact on throughput, inventory, and operating expense as a result of the decision.

Note: Using the system to make decisions based on the outcome of certain simulation attempts is a poor management strategy and should be avoided. Simulation should be used only after good decisions are made to validate them.

IX. SUMMARY

Making the day-to-day decisions for running a business can have a major impact on how well the company performs. When the results of those decisions are in line with the goal of the company profits increase. The tools provided by the system in support of the tactical decision process are an extremely important addition to the information system. However, of even more importance is the focus brought to strategic issues, such as at what point a consideration should be made to build a new factory or what will be the impact of entering a new market. Strategic issues are discussed in Chapter 11.

X. STUDY QUESTIONS

1. What is the difference between a tactical decision and a strategic decision?
2. What primary benefits are provided to purchasing through the use of the TOC-based manufacturing system?
3. When will purchasing have a conflict with the protection or creation of throughput?
4. Besides the raw material expense, a part's value varies depending on what three internal issues?
5. What minimal information should purchasing have to determine the value of a specific part?
6. When confronted with the possibility that the constraint may run out of material, how much additional expense can the vendor afford to spend to provide the raw material part?
7. What primary issues must be considered when off-loading constraint time to the vendor?
8. From the data in Fig. 41, determine which products should receive the highest commission schedule per unit and which should receive the least. Explain why.

Product	Sales Price	Raw Material	Total CCR Time
A	400	250	30
B	300	100	25
C	100	50	5
D	400	100	45

Figure 41 Data for study question 8.

9. Upon what basis is a decision made to either accept or reject an order?
10. What must be considered when direct labor becomes the constraint and how does this occur?
11. Explain the concept of throughput accounting. How does it differ from cost accounting? Give examples of each.
12. From a quality perspective, what is the most important portion of the net on which to focus to ensure the prevention of scrap?

PRODUCT	CCR TIME/UNIT	CCR TIME AVAIL
A	15	1500 MIN

ORDER = 100	SP = $200	RM. = $100

DEFECTIVE PARTS	TIME/UNIT	MATERIAL
20	15	$10

Figure 42 Data for study question 13.

13. Given the data in Fig. 42, should the decision be to scrap or rework the defective parts? Why?
14. Explain the impact that rework will have on the creation of a schedule. Give examples.
15. What is the formula for computing the effectiveness rating for total productive maintenance (TPM)? List the problems involved with its implementation and give valid solutions for overcoming the problems.

11
Making Strategic Decisions

Strategic decisions by definition will have a major impact on the company. Setting the direction for large portions of the company's assets, they include issues such as what measurement system or systems should be used to focus activity, what strategy to use in entering new markets and when to expand the factory. The decision to use the theory of constraints (TOC) including the TOC Thinking Process and TOC-based information system is definitely a strategic decision. However, for the purposes of this text, strategic decisions will be addressed in the following context:

- Market segmentation
- Company growth
- Long-term planning
- Recession proofing the company

I. MARKET SEGMENTATION

Entering new markets can be risky for any company. Many who make an attempt lose a great deal of money in the undertaking. If it were not for the entrepreneurial spirit that drives some people to take risks, new ventures would probably be rare. This kind of activity is a natural part of the business world and must continue. So reducing the risk of loss for new ventures while maximizing profit is vitally important.

The objective of market segmentation is to maximize profitability while reducing risk. Managed correctly, the market segmentation strategy can make a big difference in the success rate of new ventures as well as the survival rate of most companies.

There are numerous ways of segmenting the market, for example, by geographical region, by specific product, and by customer. However, to be successful certain rules should be followed. As presented by Dr. Goldratt, markets should be segmented so that

- The sales price in one market will not impact the sales price in another.
- The same resource base serves all markets.
- It is unlikely that all segments will be down at the same time.

Companies are not made of numerous resources acting independently, but of throughput chains. As previously defined, a throughput chain is a chain of resources connected by the process of generating throughput. In proper market segmentation, each throughput chain may feed numerous markets so that resources are protected from a downturn in any one market, or in several, so that the company can be more selective in what sales orders it takes. One of the benefits is that the company is able to manipulate the market rather than having the market manipulate the company.

A. Protecting the Sales Price

If the price of a product being sold into one market is affected by the price in another, the ability to support different pricing structures immediately falls. As an example, a manufacturer of building supplies decides to segment the market based on customer type, one being commercial buildings and the other being residential housing. The same products are sold to both markets. If the product is equally applicable in both environments, the supplier may have great difficulty in offering specific supplies to one builder at one price and to another builder at a different price. However, by modifying each part slightly to change each builder's perception of its usefulness to them, the price requested in one market may have no influence on the other. The benefits can be surprising.

In Fig. 1, products A and B are made on the same resources and are sold in completely different markets. The price charged for each has no effect on

Figure 1 Protecting the sales price.

the other. Both markets are fed by the same resources and take the same amount of time to produce. The amount of raw material used to create both is equal so the throughput generated by A is greater by $20. When the constraint is inside the company sales preference is given to A. If for some reason one market becomes soft and the asking price must be lowered to keep business, the price in both markets will not be affected. Profitability may decline, but not as much as it might have.

If a third product, product C, can be made on the same resources to feed a different market segment than products A and B, there may be even more flexibility in price. If there is normally enough volume from markets for A and B to place the constraint in the factory and for some reason market B should become depressed, a third market can be used in conjunction with A and B to continue to fill the factory with work while maintaining price protection in the original markets. When the market for B returns to its normal level, C can be given a lower sales priority depending on the amount of throughput generated per unit of the constraint.

Note: when lowering the price on B or in selling C long-term contracts should be avoided.

B. Protecting the Resources

If the same manufacturing company had two different sets of resources feeding the commercial and residential markets separately and for some reason one of the markets declines, then those resources that feed the declining market would be threatened. However, if both markets are fed by the same resources and a downturn occurred, then the threat is not as great. Figure 2 represents product flow diagrams for products A and B. Each product uses the same resources and must be processed on the constraint; they share the same throughput chain. If demand for either A or B should decline and the markets were perfectly segmented, then resources R-1 through R-3 would not be threatened.

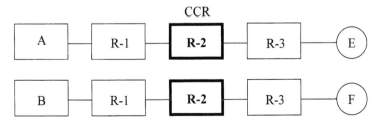

Figure 2 Protecting the resources.

Notice that the amount of throughput that can be generated by the throughput chain is limited by resource R-2. If a third market could be found which does not go through R-2, the price flexibility and resource protection would be greatly enhanced. In Fig. 3 product C represents a third market segmentation.

Product C uses resources R-1 and R-3, but avoids using resources R-2. In this case price flexibility and resource protection are increased. In addition, if resources R-1, R-3, and R-4 have excess capacity, any price charged for product C over the cost of raw material is all profit. This creates a definite competitive advantage whenever price is a consideration.

However, with the increased utilization of R-1 and R-3 comes the possibility of causing problems for the exploitation of R-2 by interfering with the subordination process of R-1 and R-3. If R-1 and R-3 begin to absorb protective capacity, then the constraint and shipping buffers would need to be enlarged, creating an increase in inventory and operating expense. If the additional load caused a secondary constraint, then inventory and operating expense would be increased even more.

The previous example does not mean that when a problem occurs in the subordination process that something is wrong, only that something must be done to protect the constraint under the increased demand. The system wil either help in scheduling through the problem or let the user know that schedules must be pushed into the future. If the schedule is pushed and the reason isolated to a specific throughput chain, then increased capacity might be warranted for a specific resource on the chain, or the sales department can limit the amount of that particular product being sold. If sales are to be limited, another throughput chain that does not have a limitation should be identified and its excess capacity sold. This

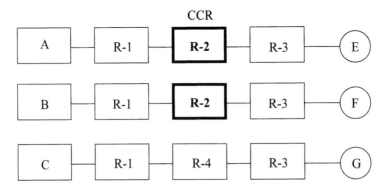

Figure 3   Price flexibility and resource protection—adding a third market segment.

process continues until all resource capacity is completely absorbed in the generation of throughput.

Identifying the throughput chains of the company, understanding what their strengths and weaknesses are, and then maximizing the market segments fed by each chain will greatly reduce the risk of failure for new market penetration while maximizing the profit generated.

C. Using the Information System in the Segmentation Strategy

Objectively, having the same resources feed more than one market where the price in one market is not affected by the other is good business. Using the TOC-based information system to aid in developing and executing an effective market segmentation strategy can offer a tremendous advantage by helping to

- Generate the strategy
- Identify the strengths and weakness of the strategy selected
- Constantly upgrade of the strategy

1. Generating the Strategy

The key issues in generating the market segmentation strategy include

- Finding the throughput chains
- Assessing the extent of market segmentation for each throughput chain
- Finding and using excess capacity to enhance pricing flexibility and the generation of throughput
- Creating and introducing a new product to further segment the market
- Understanding the impact of additional market segments on the subordination process

The key objective in identifying the throughput chains is to understand which string of resources feed what market segments.

Figure 4 is a modification of the descending part/operations screen seen previously. Throughput chain 1 feeds the market with products A, B, and C. The string of resources used to accomplish this task include R-1 through R-6. Throughput chain 2 differs from 1 in that it feeds the market with products D, E, and F; uses an additional resource (R-7); and does not include resource R-4. Notice that the constraint in throughput chain 1 has been placed in bold. Resources that have been designated as primary or secondary constraints should be identified so

Throughput Chain 1   Throughput Chain 2

PRODUCT A	PRODUCT D
PRODUCT B	PRODUCT E
PRODUCT C	PRODUCT F
R-1	R-1
R-2	R-2
R-3	R-3
R-4	R-5
R-5	R-6
R-6	R-7

Figure 4 Identifying the throughput chain.

that resources with excess capacity can be easily found. Notice that throughput chain 2 does not have a resource that has been designated as having a capacity problem.

The market segmentation screen shows each throughput chain and the extent to which it has been segmented (the number of products it feeds). It also shows which string of resources may have additional capacity that can be exploited. Once each chain has been identified, the throughput generated from each product can be compared so that the process of constantly upgrading each market segment can be accomplished. The objective is to continuously look for and identify new products that have increased throughput value.

Figure 5 shows the throughput value of each chain. Notice that throughput chain 1 is listed by throughput per unit of the constraint (uc). Since there is no resource limitation in throughput chain 2, it is listed by throughput per unit. Notice that for throughput chain number 1 the lowest value per unit of the constraint is $7.00. Selling a product below this price would cause a decrease in the amount of throughput generated.

Selling any amounts of D, E, and F would de desirable. Since there is no capacity limitation, the value of each product must be judged based on the amount of throughput per unit sold. When the throughput per unit reaches zero, the price of raw material and the sales price are equal and the product should no longer be available.

Throughput Chain 1   Throughput Chain 2

PROD.	T/uc	Total Throughput
A	$9.00	$60,000
B	$8.50	$50,000
C	$7.00	$40,000

PROD.	Throughput Per Unit	Total Throughput
D	$60.00	$55,000
E	$50.00	$45,000
F	$45.00	$40,000

Figure 5   Throughput value of the chain.

When selling excess capacity there are specific rules that apply:

- Do not oversell the market.
- Pick a market that is relatively large so that a small amount of additional product will not upset the current balance and so that it can be entered and exited without major impact.
- Avoid starting a price war.

II. GROWING THE COMPANY

Understanding how to grow a company based on its strengths is critical from a strategic perspective. It makes no sense to grow based on the weak points within the company. A key issue in the process would obviously be identifying where the weak points are located and how to address them.

Growing the company can mean entering new markets, increasing the size of the current facility, adding resources, or hiring new employees. Knowing what the strengths and weaknesses are can help considerably in accomplishing these tasks. It is important to understand that what is being advocated is not avoiding the use of the five-step process of continuous improvement or in using the thinking process to determine what to improve, but recognizing that different throughput chains will be stronger than others when maximizing growth possibilities.

As explained in *The Theory of Constraints: Applications in Quality and Manufacturing* a company should build new facilities when every bit of capacity in the old facility is being used to generate throughput and when it is expected that the new resources will have enough segmentation within the market to protect them. If a company wants to grow, it is unwise to do so where price and resources availability are critical.

Building a new facility when there is still available capacity in the old one is wasteful. As seen in market segmentation, the information system can help tremendously in finding and exploiting additional capacity. Simply implementing the five-step process from the perspective of the physical resource that has been identified as the primary constraint ignores the fact that different throughput chains will have different constraints. Most of them will not be physical. They will instead be managerial. Identifying throughput chains where additional capacity is available is an indication that, for those chains, the constraint is not physical. Since every chain must have a constraint, there must be something blocking the creation of additional throughput for that chain other than the physical resource.

Constantly growing the company through the chain that contains the primary constraint will eventually lead to a resource that is very expensive to replace. To continue to grow through this method will require a large outlay of capital. Before making this kind of commitment it would be best to ensure that every other throughput chain is being tapped for as much throughput as possible.

The strongest points within the company and the easiest way to grow is not by constantly focusing on the chain with the primary physical constraint. While it still must be exploited, by definition it is limited in its ability and will cost money to elevate. Most constraints in the company will cost nothing to elevate. Which is preferable?

Throughput Chain 1   Throughput Chain 2

PRODUCT A	PRODUCT D
PRODUCT B	PRODUCT E
PRODUCT C	PRODUCT F
R-1	R-1
R-2	R-2
R-3	R-3
R-4	R-5
R-5	R-6
R-6	R-7

Figure 6 Growing the company.

The most effective directions to grow can be found in the same fashion as discussed for market segmentation. A throughput chain whose resources have no problem in subordinating to the market or constraint schedule and do not create conflicts with other schedules is considered to have excess capacity available. The strongest points will be those throughput chains that do not go through a physical constraint.

In Fig. 6 the preferred constraint to break is not necessarily R-4. It may be the constraint for throughput chain 2.

One method for testing individual strategies would be to stimulate the decision after having reviewed each throughput chain to identify how to break its constraint. Key issues include

- The impact of the decision on other throughput chains
- The impact on the primary physical constraint within any throughput chain and where it might go next

Once individual throughput chains have been identified which do not contain resource constraints on the throughput chains list, a final check can be made by attempting to subordinate to the market demand. As an example, a decision is made to increase throughput for throughput chain 2 by adding a new product and market segment. Product H is added and a forecast entered. During the scheduling process a new constraint is created. Fortunately, there are no major conflicts which result in pushing scheduled sales orders into later

Throughput Chain 1   Throughput Chain 2

PRODUCT A	PRODUCT D
PRODUCT B	PRODUCT E
PRODUCT C	PRODUCT F
	PRODUCT H
R-1	
R-2	R-1
R-3	**R-2**
R-4	R-3
R-5	R-5
R-6	R-6
	R-7

Figure 7   Adding products to the throughput chain.

periods. The impact is registered on the throughput chains list (see Fig. 7). Notice that resource R-2 has now been identified as a secondary constraint.

While the system has no problem dealing with the schedule for the new product, there may be better choices. The result of creating a market segmentation and selling product H is that protective capacity is used. Lead times as well as inventory and operating expense will increase. If the lead time increases past the customer's tolerance lead time, new problems could be created. It may be best to first try a segment where a secondary constraint will not be created. However, it may also be possible and just as desirable to do both.

III. LONG-TERM PLANNING

In long-term planning the objectives are to understand where the company should be one to two years in the future and help guide the process. Understanding the impact that the constraint will have on the company during that period is vital. Key issues include

- Where will the constraint be one to two years from now? How does this compare to where it should be?
- What constraints will be broken along the way and when will they occur?
- What is the advantage of elevating the constraint from one location to the next?
- When the constraint is elevated will the new interaction between resources cause problems in being able to create a schedule?
- What would be the impact of building a new facility?
- If a decline in market price for a specific product is expected to continue, at what point should the product be dropped from the catalog?

The information system can help in the process by comparing different strategies for elevation over long periods. It can also help determine at what point a new resource should be purchased and what the impact to the decision process will be if it is done.

A. Elevating the Constraint

Once the constraint has been elevated the impact can be quite extensive, notice the difference between Figs. 8 and 9. If resource R-4 is the constraint, product A is the most profitable and product D the least. If resource R-6 is the constraint, however, product D is the most profitable and product C the

Resource R-4

Product	Sales Price	Raw Material	Throughput Generated	Total CCR Time	T/uc
A	300	150	150	15	$10.00
B	310	150	160	20	$8.00
C	200	100	100	15	$6.67
D	400	150	250	45	$5.56

Figure 8 Elevating the constraint. Resource R-4 is the constraint.

least. This change in the location of the constraint causes a major shift in how the company operates, how much money it is generating, and where to focus improvements. The TOC-based information system can be used as a simulation tool to determine the impact of changes that may occur months into the future, thereby providing management with the ability to see the impact of the strategic decisions.

It is important to understand that the information system assumes no policy constraints. It only deals in the physical interfaces of resources as they

Resource R-6

Product	Sales Price	Raw Material	Throughput Generated	Total CCR Time	T/uc
D	400	150	250	20	$12.50
B	310	150	160	15	$10.67
A	300	150	150	20	$7.50
C	200	100	100	30	$3.33

Figure 9 Elevating the constraint. Resource R-6 is the constraint.

are impacted by the environment in that they exist. If a physical constraint is preventing the increase of throughput, the system can deal with it. If a policy constraint exists, the system can only reflect the impact that the policy is having on the physical resources. In the long-term planning process, policy constraints are just as active as physical ones. The system assumes that they will be solved.

IV. RECESSION PROOFING

Recession proofing occurs when the company

- Continuously implements the five steps
- Minimizes the impact of a downturn in the economy by protecting resources through market segmentation
- Uses a valid decision process

The information system can help by

- Providing the basis for a valid decision process
- Managing the physical resources and scheduling the factory
- Providing a focal point for implementing the five steps
- Aiding in the identification and elimination of policy constraints
- Setting the proper resource strategy for maximizing the creation of throughput while minimizing the risk

V. SUMMARY

More than just a replacement for the cost accounting system, the TOC-based information system can offer a tremendous advantage in the strategic planning at the corporate level. It is a strategic weapon which can be used to support and guide the future growth and direction of the manufacturing company and is currently being used in numerous companies around the country.

VI. STUDY QUESTIONS

1. Explain the concept of marketing segmentation and list the three primary rules.
2. What advantages does proper marketing segmentation provide?
3. List at least four methods of segmenting the market.

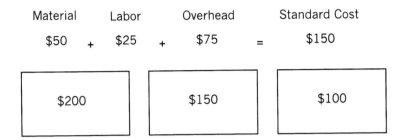

Figure 10 Data for study question 4.

4. Given the information in Fig. 10, what is the profit margin for the $200, $150, and $100 market segments? Assume excess capacity.
5. What key issues are involved in the generation of the market segmentation strategy and what can the TOC-based manufacturing system do to help?
6. Explain the advantage of using excess capacity as a weapon in the process of growing the company. What are its pitfalls?
7. How is excess capacity identified using the manufacturing system?
8. Notwithstanding regulatory issues, at what point should a company consider building a new plant?

12
Modifying the Current System

Although the tendency at this point might be to throw out the traditional system an start over, this is not a good answer to the point, the issue should be what processes and file structures should be kept, added, or eliminated. Understanding this is critical to the process of migration toward the TOC-based system.

As explained earlier, the new system is quite different from the old in a number of ways. Gone are the technologies of MRP, CRP, and the traditional shop floor scheduling methods for certain resources. Also missing is cost accounting and accounting methods such as labor and overhead allocation to work orders. Added are those processes and tools that support the five steps of improvement and throughout decision-making.

The traditional MRP-based system provides a tremendous platform on which to build. Chapter 12 will be a presentation of the end product and will describe the contents of an information system designed for the manufacturing environment.

I. CHAPTER OBJECTIVES

- To present a detailed model of the new information system
- To define what must be done to modify the traditional system
- To define the interfaces required

As seen earlier, the TOC-based information system (see Fig. 1)

- Relies on the output of the master production schedule
- Identifies the constraint by comparing those resources which have taken the most protective or productive capacity from the system
- Exploits the constraint by building a schedule and maximizing its effectiveness through setup savings, overtime, or off-loading

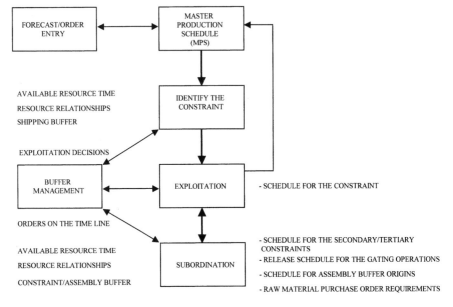

Figure 1   The TOC-based system.

- Subordinates the remaining resources to the schedule produced during the exploitation phase
- Buffers the system from those things which can and will go wrong and is used to focus on improvements, establish lead times, and determine the release schedule for raw material

These processes will require that certain data be maintained and used as input. Purchase orders and sales orders must be written, part numbers and inventory need to be maintained, money must be received, and employees as well as vendors and government must be paid. The traditional system has made great progress in supporting these areas. A very large portion of the system needs no modification at all.

II.   UNDERSTANDING WHERE CHANGE WILL OCCUR

A.   Master Production Scheduling

The traditional master production scheduling procedure will probably remain as the origination of input and definition of demand for the system. Little, if any, modification is required other than creating a link between the schedule for the constraint, the MPS file, and the net (see Fig. 2).

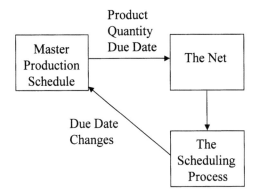

Figure 2 Linking the master production schedule.

It is the master production schedule that actually helps foster the generation of the net. After the master schedule has been set, demand is then passed from the MPS in the form of sales order/product numbers, including quantity and due dates, to the program designed to create the net. It is the MPS that determines the part/operations required to produce to the demand of the sales orders and the forecast (see Fig. 3). As will be seen, the MPS does not provide the only data that need to be passed to the net generation program.

Note: The sales order/product number is similar to the part/operation in that one string of characters is used to identify the product ordered by the customer and the sales order number.

B. Linking to the Constraint

The link between the schedule of the primary constraint and the master production schedule is made so that the MPS will accurately reflect the due

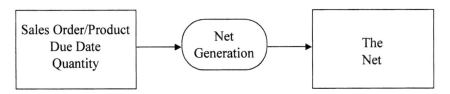

Figure 3 Linking the master production schedule.

date of the sales order or forecast. When changes occur as a result of the scheduling process the new dates are passed to the MPS file.

To update sales order and forecast due dates, the completion date and time from the constraint schedule is used. During the scheduling process the system will push those orders into the future that have rod or buffer violations. These changes are made by adding the length of the shipping buffer to the date the order is to be completed on the primary constraint (see Fig. 4).

Part/operation B/30 was originally scheduled to cross the primary constraint (R-4) by hour 8:00 on day 99. The sales order was schedule to be shipped 8 hours later according to the shipping buffer. Part/operation B/30 has been rescheduled to day 100 hour 4:30. Notice that S/O111's due date and time have also been pushed into the future by the length of the shipping buffer.

Note: if a secondary constraint has been created between the primary constraint and the sales order, then the sales order due date will be modified based on the date the order is to be completed on the secondary constraint.

To accomplish this task there must be a link created between the sales order due date from the output of the scheduling process and the master schedule (see Fig. 5).

Since the output of the net feeds directly to the MPS file, the traditional available and available to promise (ATP) data should be readily accessible (see Fig. 6).

Resource R-4

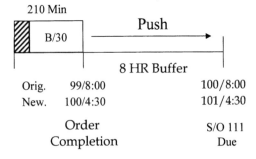

Figure 4 Pushing the order.

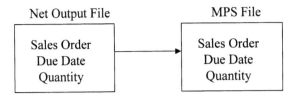

Figure 5 The link between the MPS and the net.

C. Checking Finished Goods Inventory

Since the net actually starts with the sales order/product number and the TOC scheduling process is capable of checking inventory on hand as well as demand for end items required by the sales order or the forecast, there is no need to offset sales order demand for on-hand inventory prior to placing demand data into the net.

Note: As presented in Chapter 2, rough-cut capacity planning (RCCP) will be eliminated and completely replaced by the TOC scheduling process.

D. Material and Capacity Planning

Other features of the traditional system that are to be eliminated and replaced by the TOC scheduling process include gross-to-net requirements generation (MRP) and capacity requirements planning (CRP). As the TOC scheduling process creates a schedule for the buffer origins and diverging operations, it is also generating the release to gating operations and order

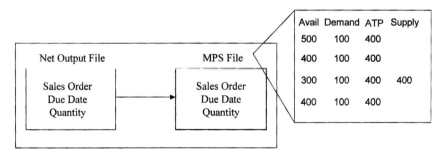

Figure 6 Available to promise.

information for raw material. TO do this the system must know what material has already been produced, what is left to produce, and the individual timing of orders. As seen in earlier chapters, the planning of material and capacity requirements is actually imbedded in the TOC logic for identification, exploitation, and subordination.

E. Action Messages

Action message processing, a standard feature in most traditional systems, can be maintained for managing purchased parts and for designating the release schedule at gating operations. Action messages include

- Order launch
- Defer
- Cancel
- Expedite

To obtain an action message the output of the TOC scheduling process should be connected to the facility used to present the message. Obviously, before the system will know what orders to defer or expedite the purchase order identity must be matched to the demand of the schedule.

F. Shop Floor Reporting

To aid in the scheduling and decision-making processes, the location of work-in-process (WIP) inventory must be known. Orders must be tracked and information fed back to the net and to buffer management so that they can work with current data. The traditional method for tracking orders at specific operations by logging in or out can still be used. Emphasis should be placed on what part/operations have been completed and where the work-in-process inventory is.

Rather than reporting every part/operation at every resource, it may be beneficial to report one part/operation and have a number of them automatically updated. The act of reporting milestones for part/operations other than those for which a schedule has been created can still be useful.

G. Labor Collection

Labor collection, a part of the traditional shop floor system used to feed cost accounting, can under normal circumstances be eliminated. Since these data will no longer be used in the decision process, hourly employees will only need to record the time they are at work for pay purposes.

H. Dispatching

A method of determining the sequencing, start times, and stop times at locations such as the constraint(s) will still be necessary. The traditional dispatch report used to sequence orders at specific locations on the shop floor (see Fig. 7) is to be replaced by the constraint or assembly schedules (see Fig. 21 later in this chapter). The schedule maintained for each operation can be eliminated for all but the buffer origins and diverging operations. The method of relating order sequencing will be by start date/time and finish date/time only. Sequencing methods such as slack time per operation and next queue should not be used.

A majority or resources will be sequenced based on the timing of material released at gating operations. Actual sequencing for those operations will be first come, first serve. When an order is received it should begin processing immediately and be completed as fast as possible.

Note: the objective of the nonconstraint schedule (see Fig. 22 later in this chapter) is slightly different. It is used to tell the operator not to start an order before a specified date and time.

I. Shop Paperwork

The paperwork necessary to process an order will be very similar to what is currently distributed with the traditional system. It should include

- A list of the parts needed for processing
- The routing
- Any special processing instructions needed at the work station

Dispatch Report R-4

Order No.	Part/Oper	QTY	Day	Start	Finish
W/O111	A/20	50	100	00:00	01:00
W/O222	D/20	30	100	01.01	01.45
W/O123	G/30	75	100	01:46	03:46
W/O333	B/10	10	100	03:47	05:00
W/O124	A/20	30	100	05:01	06:30
W/O167	A/20	40	100	06:31	07:45

Figure 7   Dispatching report.

Depending on the environment, product flow should be fairly easy to distinguish. Routings as well as parts lists may be unnecessary. However, the shop packet should include some way of distinguishing what product is being produced and what sales order or orders are to be filled.

J. Buffer Management

Buffer management will become an important part of shop floor control. A method of presenting the zones and health statistics of the buffer with current orders would be very helpful. In addition, statistics should be maintained on the buffer management worksheets and periodically consolidated into reports. There are benefits to having each of these features accessible from an interactive screen and on reports. Both should be available. Figure 8 is a screen presentation of the buffer in front of the constraint. The information to the right provides the health indicators for zones I and II of the buffer and details about the current order being processed.

Note: Zone III and the area used to pedict orders that have arrived too early are not shown due to a lack of space (see Chapter 9).

In the box to the left of the figure is a graphical representation of the orders that are supposed to be in the buffer along with the specific order numbers. Notice that all of zone I is full while 50% of zone II is full. Each column represents one hour. Notice also from the number of columns that are full that there are three hours of work in front of the constraint. Order 123 is just getting started. Current time is 1:00.

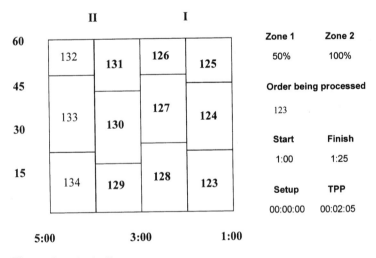

Figure 8   The buffer management screen.

The system should support the ability to display detail information about any order in the buffer. For those orders which have not reached the buffer but should be in zone I or II, the current location should be shown so that order can be found and tracked or expedited.

Figure 9 is the same detailed information as in Fig. 8. The additional data in the box belong to order 124. The part sitting in the constraint buffer is station A/30 for sales order S/OABC. There are a quantity of 10, which take 2.5 minutes each to process.

K. Accounting

The Traditional practice of allocating labor and overhead for the purposes of making decisions should be eliminated from the system. However, there will still be financial indicators. Specific data used in the TOC decision process must be made available and includes

- Raw material cost
- Sales price
- Inventory value
- The accumulated value of throughput dollar days
- The accumulated value of inventory dollar days

With the exception of throughput and inventory dollar days, the traditional system is quite capable of maintaining data on all of the above. Most

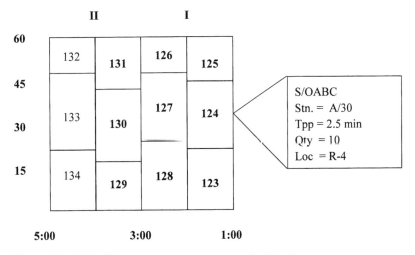

Figure 9 The buffer management screen—order detail.

$$\text{NET PROFIT} \quad = \quad \text{THROUGHPUT} \quad - \quad \text{OPERATING EXPENSE}$$

$$\text{RETURN ON INVESTMENT} \quad = \quad \frac{\text{THROUGHPUT} - \text{OPERATING EXPENSE}}{\text{INVENTORY}}$$

Figure 10 Calculating net profit and return on investment.

accounting systems are equipped with the ability to reorganize the chart of accounts for reporting purposes and can present reports showing the current and past period's throughput, inventory, and operating expense figures.

By eliminating the labor and overhead values in the item cost records and leaving only inventory cost, the gross margin (the difference between sales price and the cost of goods sold) Becomes the throughput figure. By using the output of the scheduling process to predict the orders being sold and subtracting the raw material cost, a projection can be made of the future throughput to be generated by the company. Combined with operating expense and inventory projections, familiar financial measurements can be determined for future as well as current and past periods. Figure 10 presents the formulas for computing net profit and return on investment from these data. Figure 11 presents productivity and inventory turns measurements.

Some companies will want to use the total variable cost method. The total variable cost of an order is determined by adding those costs that can be directly attributable to a given order, such as commissions and landed costs. Landed costs are those costs associated with transportation, taxes, and duties and are generally added to the cost of raw material. The result is then subtracted from the sales price to determine throughput.

$$\text{PRODUCTIVITY} \quad = \quad \frac{\text{THROUGHPUT}}{\text{OPERATING EXPENSE}}$$

$$\text{INVENTORY TURNS} \quad = \quad \frac{\text{THROUGHPUT}}{\text{INVENTORY}}$$

Figure 11 Calculating productivity and inventory turns.

L. Purchasing

The purchasing system will require no changes. The traditional system works fine for organizing, placing, and tracking orders with vendors. However, there are some other issues that need to be addressed.

Purchasing receives from the scheduling system a list of parts, quantities, and due dates. Prior to creating any action messages for placing an order, a comparison must be made with the current on-order data. At that time notices can be generated which will request that purchase orders be created, deferred, expedited, or canceled.

Additionally, as discussed earlier, purchasing needs access to the scheduling/decision system to understand the importance of certain parts when decisions need to be made, such as whether or not to pay premium to get a part flown in or whether to purchase a part that is currently made internally.

III. INTEGRATING WITH THE CURRENT SYSTEM

As earlier defined, the base for the scheduling portion of the new system is the net. It provides in one file a model of how individual resources are connected, including

- All open orders and forecasted items with the product name or number, the amount of demand for each, and the due date
- A list of all part/operations and how they are sequenced, including the time per part (run time and setup), the amount of inventory completed at each part/operation, and the resource on which they are processed
- A list of all resources and the quantity available
- All raw material, including the quantity on hand and lead time

Additional information used in the scheduling process includes the individual buffer sizes for constraint, shipping, and assembly buffers as well as a shop calendar with the work day hours and overtime availability.

Except for the buffer data and possibly allowable overtime, all the data provided to create the net currently exist in various files within the traditional system:

- All open orders and forecasts originate from the master production schedule.
- Resource information is obviously found in the resource file.
- The part/operation is a combination of the bills of material and route files.
- Raw material and the quantity on hand can be found in the inventory or item master file.

The benefit to placing all these data in one file is processing speed. For an information system to be effective it must provide immediate feedback. To answer a simple question such as what lot size to use should not take hours, only minutes or perhaps seconds. With all scheduling information stored in one file and placed in CPU memory, all disk access can be virtually eliminated from the process.

A. Input to the Net

Figure 12 illustrates the transfer of data—including the bills of material, routings, raw material inventory, resource data, and work-in-process inventory—from the traditional system to the net. Data elements from each of these files are transferred to the net to facilitate the implementation of the five steps.

B. Generating the Net

The net is generated by taking specific data from each of the sources listed in Fig. 13 and placing them into one file. The data that originated on the traditional system are used as input to the TOC-based system. Notice that the input data have been split into five different categories. Each category contains the specifics required to generate the net. In addition to the master schedule detail already discussed, the data categories include

- Raw materials
- Resources
- Part/operations
- Order of processing (to/from)

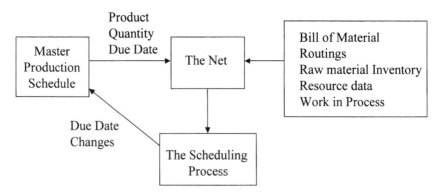

Figure 12 Input to the net.

Master Schedule	Raw Material	Resource Data	Part/ Operation	Order of Processing
S/O/Prod	RM Part	Resource	Part/Op	To - Part/Op
Quantity	Stock Qty	Quantity	Resource	From - Part/Op
Due Date	Lead Time		Run Time	Quantity
			Setup Time	Scrap
			Alt. Resource	Rework
			Run Time	Rework - Part/Op
			Setup Time	

Figure 13 Input to the net.

1. Raw Material Data

Raw material includes the raw material part numbers required to build the demand generated by the master schedule, the quantity currently in stock, and the lead time required to replenish stock (see Fig. 14).

2. Resource Data

Resource data include the resource identity of those resources used in the current production plan and the quantity or number of resources available (see Fig. 15).

Raw Material	Stock Amount	Lead Time
A	10	10
B	50	25
C	30	20
D	20	10
E	110	30

Figure 14 Raw material data.

Resource Data

Resource	Quantity
R-1	1
R-2	2
R-3	2
R-4	1
R-5	2
R-6	3
R-7	1
R-8	1

Figure 15   Resource data.

3. Part/Operation Data

Part/operation data include the part/operation to be produced, the resource where production is to occur, and the run and setup times used to produced the part at that resources (see Fig. 16).

4. Order of Processing Data

The order of processing (to/from) category designates the order in which processing is to occur. The "to part/operation" is the recipient of a com-

Part/Op	Resource	Run Time	Setup Time	Alt. Res.	Run Time	Setup Time
1234/10	R-7	1.0	1.0	R-8	1.3	1.0
1234/20	R-5	1.5	1.5	R-2	2.0	1.5
1234/30	R-4	1.0	1.0	R-6	1.5	1.0

Figure 16   Part/operation data.

pleted "from part/operation". The quantity indicates the bill of material relationship. In other words, in order to build a part/operations. In addition, a to part/operation may have more than one from part/operation reporting to it, just as in a bill of material. The from part/operation which has more than one part/operation reporting to it is called an assembly, or converging operation. "Scrap" is the percent-age figure used to increase the number of from part/operations required to produced the to part/operation based on the amount of scrap that may occur at the from part/operation. "Rework" is the percentage of rework expected, while "rework part/operation" indicates where the rework take place. In Fig. 17, part/operation 1234/10 must receive raw materials A, B, C, D, and E. There is a bill of material relationship between raw material B and part/operation 1234/10 of 3 to 1. If part/operation 1234/10 is to be reworked, it will be performed at part/operation 1234/10.

Part/operation 1234/10 feeds part/operation 1234/20 which in turn feeds part/operation 1234/30. Part/operation 1234/30 is used to fill each requirement of product F on sales order Ord 1.

For each 1234/30 that's processed, 5% are expected to be scrapped and 10% are expected to be reworked, The rework is to take place at part/operation 1234/30. The part/operation detail (Fig. 16) indicates that part/operation 1234/30 is to be processed on resource R-4.

Note: Order of processing in this case does not refer to the order of processing on the net discussed in Chapter 9.

To Part/Op	From Part/Op	Quantity	Scrap	Rework	Rework Part/Op
Ord1/F	1234/30	1	.05	.10	1234/10
1234/30	1234/20	1	.05	.10	1234/30
1234/20	1234/10	1	.05	.10	1234/20
1234/10	A	1	.10	.10	1234/10
1234/10	B	3	.10	.10	1234/10
1234/10	C	1	.00	.10	1234/10
1234/10	D	1	.00	.10	1234/10
1234/10	E	1	.00	.10	1234/10

Figure 17 Order of processing.

C. The Origination of Data for the Net

The data used to create the net originate from the traditional system. Figure 18 indicates specifically from where. Notice that all data from the master schedule and resource categories originate from either the master schedule or resource files. Raw material data are taken from the item master and inventory files, while part/operation and order of processing data originate in the bill of material and route files.

Note: Obviously, each MRP system is designed differently and, as such, may have different file structures. Data storage such as inventory may be placed in a file that manages all inventory-related data such as on hand and on order. The objective of this section is to present basis relationships between the data on the traditional system and the input requirements to the net, not to present a solution for a specific application.

In addition to the data already mentioned, the TOC-based system requires

- A horizon length to understand what orders to load onto the timeline (*Note*: It may not be necessary to load all orders to make a decision.)
- The buffer lengths for each buffer type, including shipping, constraint, and assembly, to protect the weak points within the system
- The maximum overtime that can be applied per day or week during the auto-overtime function
- The amount of artificial load, in percentage terms, to automatically add to resources which are scheduled to a maximum capacity for a set number of consecutive days
- A shop day calendar indicating the work hours per day and the work days available during the year

These data are actually stored outside the traditional system files but are used the net.

Master Schedule	Raw Material	Resource Data	Part/ Operation	Order of Processing
Master Schedule	Item Master	Resource	Bill of Material	Bill of Material
	Inventory		Routing	Routing

Figure 18   The origination of data for the net.

D. Output of the Net

The output of the net can be found in Fig. 19. Notice that the output includes the schedule for each buffer origin, the release schedule to the gating operations, a schedule for those resources which occur immediately following the diverging, the raw material requirements for purchasing, and the overtime requirements for production.

   Note: The schedule for resources that process parts immediately following a diverging operation are scheduled to insure that part/operations are not processed before a specific time and are processed only a specific amount, nothing more.

   Figure 20 is a detailed list of those data provided after all three steps (identification, exploitation, and subordination) of the scheduling process have been completed.

1. The Constraint Schedule

Output from the constraint(s) schedule includes the identification of the constraint resources, the identification of the orders which must cross it, the order/product involved, the quantity of each, the run time, the setup time, the start time and completion date of each order, and the original completion date and time (see Fig. 21).

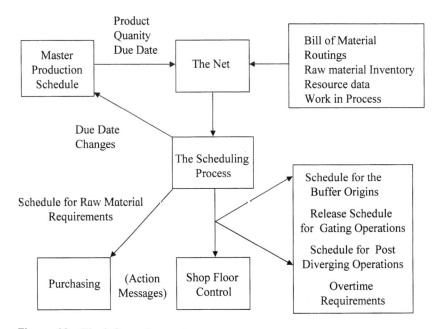

Figure 19   The information system.

Constraint(s) Schedule	Non-Constraint Schedule	Release Schedule	Raw Mat. Schedule	Sales Order Schedule	Overtime Schedule
Constraint ID	Resource ID	Resource ID	RM Part	Sales Order	Resource
Order ID	Order ID	Part/Op	Stock Qty	Due Date	Date
Part/Op	Start Date/Time	Quantity	Order Date	Quantity	Amount
Order/Prod	Due Date/Time	Start Date/Time	Delivery Date		
Quantity			Order Number		
Run Time					
Setup Time					
Start Date/Time					
Finish Date/Time					
Orig. Date/Time					

Figure 20   Output from the scheduling process.

Note: Order ID refers to an internal designation for the identification of a specific part operation on the constraint schedule.

2.   The Nonconstraint Schedule

The nonconstraint schedule represents, in detail, a schedule used to prevent certain action from taking place before a specific time (see Fig. 22). Notice that the start and stop times for each order are not consecutive. In other

Const. ID	Order ID	Part/Op	Order/ Prod	Qty	Run Time	Setup Time	Start Date/ Time	Finish Date/ Time	Orig Date/ Time
R-4	1234	A/30	S/O111/X	50	50.:00	1:00	100/1:00	106/4:00	106/8:00
R-4	1235	H/30	S/O333/Y	20	20.:00	1:50	106/4:01	108/5:50	108/8:00
R-4	1236	G/30	S/O222/Z	30	30.:00	2:00	108/5:51	112/5:51	113/8:00
R-4	1238	N/20	S/O111/Q	60	30.:00	1:00	112/5:52	116/4:52	116/8:00
R-4	1239	F/30	S/O444/R	10	20.:00	1:50	116/4:53	119/0:50	118/8:00

Figure 21   The constraint schedule.

Res. ID.	Order ID	Start Date/ Time	Due Date/ Time
R-1	Ord/123	106/4:30	107/3:30
R-1	Ord/234	108/2:30	109/3:30
R-1	Ord/245	110/4:30	111/2:30
R-1	Ord/234	116/2:30	116/7:30

Figure 22 The nonconstraint schedule.

words, when order Ord/123 is complete at day 107 hour 3:30, order Ord/234 does not begin immediately. In fact, it does not start until almost one day later. This does not mean that resource R-1 does not have any work other than what is shown on the schedule. It only means that for some reason the system does not want these order to begin before specific date and time. The reason may be that resource R-1 has inventory that it could process, but because this inventory is to be used at another resource, R-1 is told not to process until a specific time. By processing the inventory before the designated time, R-1 could be taking material that was intended for other purposes.

3.   The Release Schedule

Another output of the system is the release of schedule for the gating operations. In Fig. 23, R-7 is to begin processing part/operation 123/10 for a quality of 100 at day hour 3:30. This tells production control when to release the order and when material should be presented to the resource

Res. ID.	Part/Op.	Qty	Start Date/ Time
R-7	123/10	100	100/3:30
R-7	234/10	150	101/2:30
R-8	245/10	200	101/2:30
R-8	234/10	100	101/7:30

Figure 23 The release schedule.

Res. ID.	Part/Op.	Qty	Start Date/ Time
R-7	123/10	100	100/3:30
R-7	234/10	150	101/2:30
R-8	245/10	200	101/2:30
R-8	234/10	100	101/7:30

Raw Mat.
A
B
C
D
E

Figure 24   The release schedule with raw material data.

R-7 begin production. Part/operation 234/10 is to be release a day later at hour 2:30.

In addition to understanding when to begin processing a specific part/ operation, it is also important to understand what raw will be required to begin the process. The order of processing data within the net should indicate this (see Fig. 24).

4.   The Raw Material Schedule

The raw material schedule indicates what raw material needs to be purchased, when the order should be placed, and when to schedule delivery. In addition, it makes direct connection to the order involved (see Fig. 25).

RM Part	Order Qty	Order Date	Del. Date	Sales Order
Gasket	100	100	107	S/O111/X
Bolt	150	101	107	S/O333/Y
Battery	200	101	107	S/O222/Z
Pipe	100	101	108	S/O222/Z

Figure 25   The raw material schedule.

Order/ Prod	Qty	Due Date	Past Due
S/O111/X	10	90	N
S/O333/Y	15	91	Y
S/O222/Z	20	91	N
S/O222/Z	10	91	N

Figure 26   Sales order schedule.

5. Sales Orders Schedule

The sales order schedule indicates the specific sales order, when it is due, and the quantity to be shipped (see Fig. 26). Also indication of whether the order is late or on time based on the extent of penetration into the buffer. Notice that sales order S/0333 for product Y at a quantity of 15 is due on shop day 90. It is also indicating that because S/0333 has penetrated over 50% of the shipping buffer it will be late.

6. Overtime Requirements Schedule

When generating the schedule, a decision must be made whether or not to use the manual or auto-overtime features discussed earlier. If the feature is used and overtime is added, what resource or resources will receive the overtime and when? The overtime requirements schedule is designated to answer these questions. It lists the resource on which the overtime is to be performed, the date, and the amount of overtime required by schedule (see Fig. 27).

Resource	Date	Amount
R-4	97	2:00
R-4	98	1:50
R-4	101	2:00
R-4	102	1:00

Figure 27   The overtime requirements schedule.

Resource R-4 has been scheduled overtime on shop days 97, 98, 101, and 102. The amount of time required by the schedule is 2 hours, 1 hour and 50 minutes, 2 hours, and 1 hour, respectively. These data can be readily used to compute the increase in operating expense necessary to support the schedule as well as plan labor requirements during the schedule horizon.

E. Software Development

Integration and implementation of any information system will require the creation of at least four environments within the computer:

- Development
- Testing
- Conference room pilot
- Production

Modifying the traditional system should be no different. The objective is to protect each environment so that as the project progresses it can be controlled. Programs that have been developed by the programming staff should not be placed directly into production without being tested first. Individual programs and modules are tested by applications specialists in the testing environment once programming is complete. If a problem occurs, the programmer is notified and the problem resolved.

While the system is being developed and tested, procedures are written to explain to the end user how the system is operate. The conference room pilot is used to test the system and the written procedures in their entirety once the integration process is complete. Upon completion, the software with all tested modifications is placed into production. Production is where the actual end user of the system becomes involved.

There is currently a fledgling industry of small software development companies that have TOC-based software that can be readily integrated into most standard MRP II type systems. This will help considerably in the process and, since the preferred method of structuring the system may be client/server, the PC-based system may be the best route to take.

If for some reason the size of the company prevents the use of a PC-based scheduling system (industries such as aircraft manufacturing will have part/operations in the millions, then the more traditional integration process may be necessary. However, starting from scratch without the benefit of experience in the development of TOC-based software could be a nightmare. It may be best to take an existing system and port it to another platform.

1. The Platform

There are two issues as to where the files and logic of the system should be placed to maximize its effectiveness given the current computer technology:

- If the logic is to be placed on the host, then it must be capable of processing multiple schedules simultaneously as people are attempting to make decisions. This could seriously disrupt the response time for other tasks that must be performed.
- If it is not to be placed on the personal computer, then everyone would be able to process their decisions independently without impacting the response time. However, if the files and schedules involved in the decision process are not duplicated on each machine, then each person may approach the decision process differently.

As an example, production control sets a schedule for the factory and engineering must make a decision on how money is to be spent for improvements. Engineering then creates a separate schedule. During the process, engineering makes different decisions while exploiting the constraint and performing subordination, or they start the process with different data residing in the net. the resulting decision may be different than what is actually required. Engineering may use money to fix the wrong resource. Unless the factory and the decision process are in sync the decision process will be constantly variable and will not be based on the reality that currently exists on the shop floor.

It may be that the best environment for this technology is client/server. There are some very convincing arguments for this position:

- If the net were to exist and be maintained on a server, as changes happen within the factory a data for repository would exist which would be consistent for everyone without placing the burden on the host.
- If the logic for scheduling, subordination, and buffer management were on the server as well, then any time a need arises to have one primary schedule it could be easily done.
- If the logic were to be duplicated on the client and an ability existed to pull the net from the server into CPU memory on the client, then the independent decisions could be supported without placing the burden on the server.

Figure 28 provides an illustration of the client/server organization of the system. The host would maintain the data repository for primary file maintenance including sales orders, purchase orders, and the item master. The server would maintain

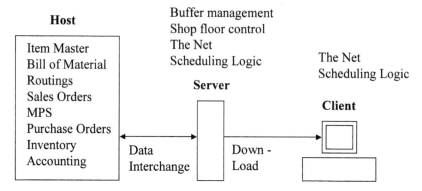

Figure 28   The client/server environment.

- The buffer management system
- Shop floor control
- The net
- Scheduling logic

The data necessary to maintain the net on the server or the data files on the host can be downloaded and uploaded at specific intervals during the day. The client would maintain a copy of the logic necessary to support the decision process. Any time a decision is to be made at any level in the company a current copy of the net can be pulled into the client for processing.

An alternative to this configuration would be to go directly to the host, download the requisite files, and generate the net at the PC level for each department. To prevent accidental modifications of the primary net, separate

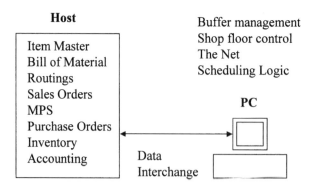

Figure 29   Interfacing directly to the PC.

files can be maintained for different simulations as well as the primary schedule (see Fig. 29)

The advantages to using Client/server are

- The host is free to maintain communication to all users for maintaining data without being burdened by the scheduling or decision processes.
- The scheduling process can be performed at any time.
- Users can make decisions any time.

IV. SUMMARY

The nature of the information system continues to improve as more and more companies adopt the theory of constraints. It is important for current information systems developers to understand where the traditional system should be modified and how it can be accomplished.

V. STUDY QUESTIONS

1. In the transition from the MRP-based system to the TOC-based system, what features will more than likely be eliminated and what features kept?
2. What data are required to build the net and from where will it originated on the traditional system?
3. What are the minimum four stages of software development or integration that should be used during the technical phase of implementation?
4. What steps within the five-step process will contain at least a portion of the logic used to plan material requirements?
5. What steps within the five-step process will contain at least a portion of the logic used to generate the schedule?
6. What portion of the system will be used to replace capacity requirements planning?
7. Present at least three examples of how the system will be used to replace cost accounting. Include a decision model in the examples.
8. Explain the concept of total variable cost and how it should be used in the system to make decisions.
9. Explain in detail the physical laws that are involved in the creation of the dynamic buffering process. (Research question.)
10. What is the objective of the order of processing category of data taken from the MRP data files?

11. What is the objective of the non-constraint schedule?
12. What data input requirements are needed to feed back to the system through stop floor control?
13. What part/operation data are required to generate the schedule and where does it originate on the host manufacturing system?
14. What data are required to generate the schedule and where does it originate on the host manufacturing system?
15. What data are contained in the following schedules and what are they used for?

    - The release schedule
    - The raw materials schedule
    - The sales order schedule
    - The overtime requirements schedule

13
Implementing a TOC-Based Information System

Implementing a different kind of system will mean using a different approach. This chapter presents the basic structure of how a TOC-based information system should be implemented. The key issue is to ensure that when the implementation is complete, the company will be able to use the system to attain its goal. This means that not only should the system be available and work properly, but also that it should be used to support the implementation of the theory of constraints.

I. CHAPTER OBJECTIVES

- To present a method of implementation for the TOC-based information system
- To discuss the obstacles which may block a successful implementation and how to deal with them

II. DEFINING A SUCCESSFUL IMPLEMENTATION

Implementing a TOC-based system represents a change from the traditional approach for a number of reasons:

- The logic of the system is very different and will require a total change in how the planning process is executed.
- It means implementing the theory of constraints at the same time. This alone is a monumental task.
- It requires a change in how the factory is organized, from either a work order, rate-based, or just-in-time environment to the drum-buffer-rope process.

- It changes drastically the role of the information system in the decision process.

The objective of the implementation of a TOC-based information system is to provide a tool to be used by the company in reaching its goal.

The implementation will be successful when those who must use the system in the implementation of the five steps can do so with predictable results, and when the results are in line with the goal of the corporation.

Predictable results can be obtained when

- A road map exists which defines the goal of the company and how to get there.
- A reliable process exists in a stable environment which will give the same results under the same conditions every time.
- A method exists to quantify the result of a specific action or provide evidence that a specific action will produce a quantifiable result.
- The process is not blocked during execution.

III. THE IMPLEMENTATION PROCESS

Since each implementation involves more than simply providing a scheduling platform, the discussion has been split into three parts with each topic having significant importance:

1. The technical implementation
2. The physical implementation
3. The logical implementation

A. The Technical Implementation

The technical implementation involves preparing the system so that it can perform the scheduling function. It includes

- Selecting the platform
- Creating the interface
- Preparing the net
- Security
- Testing the logic

For our purposes, we will assume that a traditional manufacturing system currently exists and that data management processes—such as part master data and bills of material management, sales order processing, purchasing and inventory management. etc.—have been implemented and that they are functioning correctly.

1. Selecting the Platform

Obviously, the selection of the platform on which the system is to be run cannot be determined without understanding some key issues relating to the vendor or designer for the specific application. However, there are some guidelines that can be used to ensure maximum performance from the users perspective. Key issues include

- Central processing unit (CPU) speed
- The amount of internal (CPU) memory
- Disk space

If the objective is to use the system to make decisions, then the speed at which data are manipulated must be such that any decision can be made in less than 10–15 minutes. That is the purpose of shrinking the total size of the data to be manipulated to the size of one file and placing it in internal memory. By limiting the amount of disk access requirements, the TOC-based system takes advantage of the advances being made in computer processor speeds. Using internal memory can increase the rate at which data are manipulated 10–20,000 times and, according to IBM, the increase in processing speeds will climb exponentially over the next few years.

While the speed at which the data are manipulated is a function of the processor used, the size of the internal memory is directly related to the size of the net. In configuring the system, it is important to know the maximum amount of data that will be stored in memory without having to go to the disk. Since the net is comprised of part/operations, there will be a direct correlation between the number of part/operations to be stored and the size of the internal CPU memory required. The number of part/operations within the net can be predicted by multiplying the number of sales order line items plus the forecast by the average levels within the bills of material by the average number of routing steps (see Fig. 1). In the figure, there are 1000 line items between the sales orders on file and the forecast generated for

Sales Orders/ Forecast		Levels of BOM		Route Steps
1,000	x	5	x	7

Part/Operations = 35,000

Figure 1 Determining the number of part/operations.

the company; the average number of levels in the bills of material file is five; and there are an average of seven routing steps for each assembled item in the route file. In this case, the system should be sized to handle a total of 35,000 part/operations in the net.

2. Creating the Interface and Preparing the Net

Systems currently available to support TOC-based systems on a PC will include import tools. Import tools are software-driven tools that will read specific data files downloaded form the host manufacturing system to the server or PC, consolidate the data, and write the file contents to the net file.

Detail files will include the data discussed in Chapter 12, such as

- The order in which part/operations are processed
- Available inventory
- Resources
- Overtime availability
- Master schedule data

If this tool does not exist, it will need to be written. Chapter 12 presents in detail what data must be included in the net and from where it originates in the traditional manufacturing system.

a. Refreshing the Net. How often the net needs to be refreshed depends on how dynamic the environment is. If significant changes are being made in the bill of material, routing, or master schedule during the day and those changes need to be reflected in the schedule, then an update of the net on the server should be made in proportion to the rate of change that occurs. In the client/server environment shop floor data are being maintained at the server level. Thus, refreshing the net from the host as changes occur on the shop floor will not be necessary. It is only when changes occur on the host that net must be refreshed from the host.

Changes occurring on the host manufacturing system that will result in a need to refresh the net are

- Old sales orders being filled and new ones added
- Changes in the forecast
- Bills of materials and routings being updated
- New resources added and old ones deleted
- Changes in the due date of an order originating outside the system
- Significant changes in lead times for raw material
- Changes in setup or run time for those products which are processed on the constraint or on near-constraint resources

Note: An assumption is being made that the changes being made on the host will impact orders inside the time horizon for the schedule.

3. System Security

Other than the normal security on the host system, there are three areas in which security will be a key issue:

1. Making direct changes to the primary net on the server
2. Creating and making changes to the schedule
3. Shop floor reporting

The data residing on the server are the basis from which the net is imported to the client to support the decision process. It is important that these data remain the same for all who need access to make a decision. If for any reason the base net should be compromised, people could be making decisions based on different input. As an example, suppose that engineering were making a decision on buying a specific resource and increased the quantity of that resource on the net. When production control needs to schedule the factory, the environment will not resemble the current factory. Obviously, this is not an acceptable situation. There must be a method of password securing the primary net on the server so that only a few people have access to be able to add to, change, or delete from it.

The capability to copy the net and create a new version, which could be password secured by the individual user, is also desirable. Users can thus examine several scenarios in which different shop floor environments are being used at the same time. Having access to numerous versions would save time by not having to constantly modify the same scenario over and over again to meet new changes.

Understanding where inventory is on the shop floor is necessary to effectively manage the shop and is also necessary as input to the scheduling process. Accuracy is critical. Access to material movement on the primary net should be password secured and limited to only those who are responsible for insuring that material is where it is supposed to be.

4. Testing the Logic

In testing the system a subset of data should be used in a controlled environment to ensure that the system functions as desired. Once the interface is complete, choose specific data to enter and test. Make sure that the data selected for the test are representative of all testing scenarios to be performed. Start with the creation of the master schedule. Place one order

Descending

S/O111 - A

A/30   R-1

A/20   R-2

A/10   R-3

B/30   **R-4**

B/20   R-5

B/10   R-6

RM C

Figure 2   Descending part/operation chart.

on the MPS for a small amount so that it can be tracked through the creation of the net. Look at the descending part/operation chart (see Fig. 2).

Verify that every part/operation that is supposed to be in the net is there, that they appear in the proper sequence, and that each setup and run time is correct according to the data placed in the route file. Then place the orders on the timeline by creating the schedule. There will be only one, perhaps two, orders on any one resource, but it will be important to ensure that they appear on the timeline in the proper position given the buffer length for shipping and that functions such as off-loading to the other resources works. Each order on the time line should be identified along with the setup and run times for each part/operation. Ensure that each resource can also be identified along with specific resource data.

During testing and in scheduling the factory, the system should permit manually identifying the constraint to force a schedule of a specific resource. Once the initial phase of testing is complete, expand the test to include a larger data set and manually identify the constraint. Check that each part/operation crossing the constraint has a start and stop time and that all rod types are in place and functioning. Figure 3 shows a feature that can be used to gain specific information about orders on the timeline. Notice that the applicable sales order, quantity, setup times, run times, and start stop times appear. Make sure that this data are correct based on manual mathematical computations.

a. The Conference Room Pilot.   The conference room pilot is the final test software and operating procedures prior to the physical

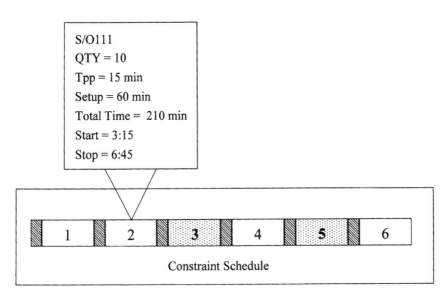

Figure 3 Testing the schedule.

implementation. It should be used to test the ability of the software and procedures to accomplish the task they were designed to perform and should not be used as a test bed for MIS to ensure that the software actually works.

Every possible scenario should be tested, from scheduling the factory to making decisions. Although a tremendous amount of time will be spent testing, it is literally impossible to predict every situation. The number of issues that will emanate from the conference room pilot which were not predicted prior to developing the operating procedures will be amazing. That is one of the benefits in using the TOC thinking process discussed later in this chapter.

Included in the conference room pilot should be tests for specific functionality such as

- Generating the schedule for the constraint
- Looping back to the master schedule
- Handling rework and scrap
- Subordinating the activities of departments such as engineering, maintenance, and quality
- Managing the buffer by reviewing its health indicators, controlling the buffer lengths, and expediting past due orders
- Decision-making

- Moving material and entering data
- Focusing on improvements

The conference room pilot is absolutely essential in the success or failure of any system implementation. When attempting the kind of changes required by the theory of constraints, it becomes mandatory to test every aspect of the system prior to implementation.

B.   The Physical Implementation

The physical implementation involves the interface with the users and the user's environment. It includes

- Modeling the plant
- Creating and following the schedule
- Buffer management

This begins the actual process of using the system to schedule the factory, focus the improvement process, and make decisions.

1.   Modeling the Plant

The first step in the physical implementation is model the plant. Modeling the plant means using the live data from the active host system to generate the net and to schedule the factory. The plant model is a view of how the facility might look if no policy constraints existed. The objective is not to use the data to sequence material flow within the factory, but to determine where the constraints are, where they should be, and what must be done, if anything, to arrange the shape of the factory, such as placing and locating of physical buffer areas.

The plant model also provides an excellent opportunity to view the impact of current non–TOC-based policies on the profitability of the company. For example, what would be the impact if

- The correct products were being sold into the market
- The constraint was being exploited internally by maximizing its utilization
- Physical improvements were made prior to implementation which resulted in eliminating irresolvable conflicts between primary and secondary constraints

Modeling the plant is an extremely important step in the implementation of the TOC-based system. It can provide some insight into the way things ought to be and can serve as performance target.

2. The Shop Floor Layout

A major change will occur in the way the factory appears. Figure 4 presents the suggested changes to the layout in which physical buffer areas are created based on certain characteristics. Notice that buffer areas have been prepared for the constraint, shipping, and assembly buffers. Since additional time is provided to allow material to arrive from the gating operations, orders will tend to collect in these areas. The shop should be prepared for this event and have specific space reserved. In addition, if enough material collects in these areas, a system may be needed to keep track of it. Specific attention should be paid to ensure that material to be processed on the buffer origin is readily available.

Once the conference room pilot has been completed and all procedures establish for the use of the system, it is time to bring the implementation to the shop floor. Since all other process such as buffer management and focusing on improvements are dependent on the schedule, the primary issue will be to begin the process of identifying and scheduling the constraint. Prior to creating the schedule a trial buffer size should be used which is initially oversized. It will be obvious to buffer management that the buffer is too large. However, starting with an oversized buffer will ensure that the constraint is well protected when the process begins. The rule of thumb for the initial buffer length is three times the normal lead time. Once the system is stabilized and results become fairly predictable, begin shrinking the size of the buffers so that zones I and II reflect the desired rates.

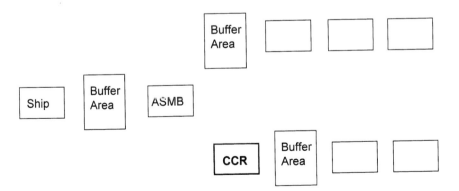

Figure 4 Suggested changes to the plant layout.

It is important that the system reflects what is actually happening on the shop floor. Begin moving and reporting material between part/operations as it is completed.

Just because the system identifies a specific resource as the primary constraint does not mean that it is. There may be data accuracy issues that are preventing the system from processing correctly. Verify that the system is reporting what is actually happening on the shop floor. If material is collecting at a resource other than a constraint or the resource that has been identified as having limited capacity has no problem in keeping up with demand from other resources, begin investigating whether the data on these two resources are accurate. Since the system will permit the manual identification of the constraint, it its possible to establish a schedule for a resource that is giving physical signs of being the constraint, regardless of what the data may indicate.

3. Shrinking the Size of the Transfer Batch

Companies using the repetitive manufacturing process or that have implemented just-in-time manufacturing systems have already been through the process of shrinking the transfer batch size. The best transfer batch size is a quantity of one. This is not to be confused with the process batch size. The best process batch size is one that maximizes utilization of the constraint. Most companies do not operate on this concept.

If the company is not currently operating with a transfer batch of one and it has been decided to make this transition, allow time during the implementation to master this technique. One of the toughest things to remember is that it is the buffer origins which must not run out of material. As inventories begin to fall and productivity increases, there will be an overwhelming urge on the part of some people who work at nonconstraint resources to stay busy or to build ahead. The logical implementation will help tremendously.

C. The Logical Implementation

With 90 to 95% of the constraints to a company's making more money being other than physical limitations, the probability is very high that after the system has been implemented the constraint will not have been broken. Nor does it guarantee that it will not be broken in the future. The system will indicate that a physical constraint either does or does not exist and provide the basis for a valid decision process. However, even when these things are provided, it docs not mean that a company will be guaranteed to make more money. Consistently implementing the five steps of improvement will.

Assuming that the technical and physical implementations have been successful, the primary issues involving implementation become

- The TOC decision process
- Dealing with the human interaction that will either enhance or block the successful implementation of the TOC-based system
- Developing a successful implementation plan to support the implementation process

1. Overcoming Blocking Actions

Blocking occurs when a member of the company acts counter to the objective of the system. As an example, if a schedule is created on a resource that follows a diverging operation and a worker decides to work ahead in the schedule processing material that was to go to another part operation, then the actions of the worker will block the success of the company in maximizing its goal by mismatching parts.

Ensuring the success of the implementation requires educating people to inform them of the impact of their actions. It also requires a measurement system that will ensure that when blocking actions occur pressure will be exerted to correct the situation.

2. Education

Oliver Wight was once quoted as saying, "If you think education is expensive you should try ignorance for a while." With a TOC-based system people are more involved in the actual process of generating the schedule than with the traditional system. The usual procedure for MRP and CRP is to run a batch program and then generate a report or action messages when the system has finished its processing. The education process consists of defining what MRP or CRP are and how the system fits together. With a TOC-based system the user is very much an active participant, making decisions at each step during the exploitation or subordination procedure. The user must answer questions such as

- Should setup savings be performed on a certain resource or should sales orders be pushed into the future? If the latter, which orders?
- Should overtime be applied? If so, where?
- What is the most profitable product mix given the current situation?
- Should the part be made on internal resources or purchased from an outside vendor?

This means that education is even more important in making the TOC system work. It's not so much what buttons to push that needs answering, but questions of a more dynamic nature. If the decision process is to change to the extent suggested and if the information system is to play a major role, then knowing how to make decisions using the information system also becomes an important factor.

What must be overcome is 50 to 100 years of an education and industrial system plagued by the almost dogmatic tendency to view every activity as an outlet of cost that must be improved. This one issue can destroy the implementation of the TOC-based system. Education is crucial.

Because something is difficult does not mean it cannot or should not be done. Everyone who has spent a short time around the theory of constraints realizes that it's the only viable solution. Avoiding its implementation is like going to the doctor, being diagnosed with hepatitis, and then not taking the medicine prescribed.

The first step in the implementation of the system must be to educate the people who use it either directly or indirectly. This includes anyone involved in the decision process, whether tactical or strategic, and anyone who must act on the decisions made by others. If production control produces a schedule and a foreman decides to maximize the utilization of all resources on the shop floor, then the schedule becomes useless. The foreman must understand the impact of what their actions are to the rest of the company.

Another example of a problem that can be created by the uninformed is the impact of irrational decisions with respect to market segmentation. If the markets fed by a specific resource chain are appropriately segmented and the chief financial officer decides that the company's resources should be split down product lines to more effectively manage cost, the segmentation process would be completely destroyed, leaving the resource base exposed to market fluctuations.

Education of all employees within the company is imperative. In addition, after the process is complete the people must take ownership of the information transferred. There are two issues:

1. The education process must be of such a nature that it causes people to change of their own volition.
2. The education process must be followed by a change in the measurement system. If the measurement system is not changed, then people will not change. They will continue to maximize the measurements they are given.

Note: A good resource for understanding the measurement systems and the changes required can be found in Lockamy and Cox's *Re-Engineering Performance Measurement* (1994).

In the case where the constraint is constantly being elevated, there is a good chance that eventually it will arrive at the office of someone who does not understand the impact of implementing the five steps or who may even be hostile to the prospect of leaving the cost world methodologies. Many arguments have resulted from the lack of understanding on the part of key people within numerous companies. These people must be educated immediately so that the process does not stop. In many cases, the emphasis necessary to convince some people listen will come from the office of the president.

a. The Permanent Education Program. As people leave the company and are replaced, new employees will be coming from environments where cost accounting is the primary driver. It is amazing how many people are still uninformed. Even people who have made the transition from the cost world will need refresher training or training in other areas as their careers progress. This requires that a permanent education program be established.

Educational requirements for the implementation of the TOC-based information system include

- The cost world versus the throughput world
- The concept of "continuous profit improvement"
- Understanding the TOC decision process
- The drum-buffer-rope (DBR) concept
- The information system
- The TOC thinking process

Examination of the cost world versus throughput world is designed to bring users to the obvious conclusion that a change must be made in how companies are run and in the decision process. Understanding the decision process is more complex and may need to be explained in the context of the job to be performed. Training the shop floor supervisor how to make sales decisions may be a waste of time.

As presented in Chapter 13 of *The Theory of Constraints: Applications in Quality and Manufacturing* (1997), specific educational requirements for the implementation of the DBR process include

- How resources interface
- How a schedule should be generated
- What the rules of resource activation are
- How material releases are determined
- What the five steps are in the process of continous improvement
- How to exploit the constraint

- How to properly subordinate
- How to make decisions

A list of these kinds of issues should be created for each departmental function for the company and education sessions performed.

b. Simulation as a Training Tool. When making this kind of a change, one of the best methods is to place the individual in the environment in which he or she must perform and present the outcome of the decisions made. This process is best done through simulation. The simulator can be used to educate people in a number of different areas, including

- Quality
- Cost accounting
- Resource Maintenance
- Setup reduction
- Purchasing
- Product design
- Market segmentation
- Growing the company

Figure 5 is representative of how a simulator for the manufacturing environment might appear. Each resource type is represented by a specific pattern. Notice that resource types R-2, R-3, and R-4 have a quantity of two.

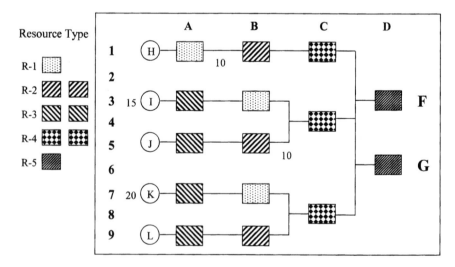

Figure 5 The education simulator.

Each resource is connected from the gating operation to the sales order through specific part/operations. Raw material enters the process from the left and appears in circles. Finished products F and G leave in the process from the right. A specific part/operation is found by an intersection of the numbers appearing down the left side and the letters appearing across the top. The last part/operation before product F is shipped is 3/D and is processed on resource R-5. Notice also that certain part/operations have inventory. Part/operation 7/A has 20 Ks ready to process, while 1/B has ten 1/As to process. Part/operation 4/C has only some of its parts required to begin processing. It has ten 5/Bs and need ten 3/Bs as well. The objective of this screen is to simulate the decisions made by the user and to predict an outcome based on the physical interfaces and characteristics of each resource.

Depending on the decisions to be made additional information will be needed. Figure 6A is list of the resources and their available statistics. Notice that the scrap rate for R-3 is 15%, that R-1 has scrap rate of 3%, and that downtime and rework are also listed.

To complete the picture will require data on the amount of setup and run time for a specific part/operation and the demand placed on the system through sales orders (see Figure 6B). Now all the data are available, the student can be led through different methodologies focusing on the different processes, such as total productive maintenance or single minute exchange of die. Decision processes such as make/buy or scrap/rework can be worked out just as easily.

When making the decision to take a specific action such as reducing the amount of rework at resource R-1 or R-4, the student will see the impact of the decision on throughput, inventory, and operating expense after running the simulation and looking at the final profitability statistics (see Fig. 7).

The student can make repairs to specific resources or products based on cost accounting, then use TOC-based throughput accounting. Different actions will result in an increase or decrease to throughput, inventory, or operating expense. To repair a resource causes operating expense to increase and, therefore, there must also be an increase in the throughput or the company loses money.

This kind of an educational tool is extremely valuable in developing an education program designed to change the perspective of those who must run the company on a day-to-day basis. It should be available in every TOC-managed company and considered part of the requirements for a permanent education program.

   c.   *Using the TOC-Based System as a Training Aid.*   After completing the conceptual education, the next phase of education should include

Resources	Time Available	Down Time	Scrap Rate	Rework Rate
R-1	480	0%	3%	5%
R-2	960	10%	0%	0%
R-3	960	0%	15%	0%
R-4	480	0%	0%	15%
R-5	480	25%	0%	0%

(A)

Part/Op.	Setup	Run
1/A	30	10
3/A	15	3
5/A	60	5
7/A	30	10
9/A	100	5
1/B	15	2
3/B	30	15
5/B	60	5
7/B	45	10
9/B	50	5

Part/Op.	Setup	Run
1/C	30	5
4/C	60	10
8/C	60	5
3/D	30	3
6/D	15	12

Product	QTY	Due
F	100	251
G	150	251
F	50	254
G	50	255
F	100	255
F	100	256

(B)

Figure 6 (A) Resource data; (B) part/operation data.

		Beginning	Ending
		Beginning	Ending

Throughput: $32,000   Inventory: $63,000 $57,000

Operating Expense: $10,000

Return On Investment: .20

Net Profit: $12,000

Productivity: 3.2

Inventory Turns: .53

Figure 7   The bottom line results of the decision after running the simulator.

training on the TOC-based system by placing the student in the actual environment in which he or she is to perform. Using a copy or subset of the actual environment as a training aid is excellent for insuring that the student can move from the conceptual world to reality.

In generating the course curriculum the instructor should develop specific scenarios involving the decision process for each functional department. Each scenario would involve a specific decision under a given set of circumstances, such as

- Establishing the correct product mix
- Making or buying a certain product
- Generating the schedule
- Setup reduction

Under the certain circumstances increasing the productivity of a specific resource may cause more problems in being able to resolve conflict between schedules on the primary and secondary constraint. Properly presented, the impact of decisions on the schedule will be readily seen, and the student will understand the impact of individual decisions on the three measurements of throughput inventory and operating expense

3.   The Measurement System

As mentioned earlier, education must be followed by a change in the measurement system to reinforce the desire to act in a manner consistent with the goal of the corporation. If an action such as releasing or processing material earlier than required by the schedule will result in a negative impact on the company by increasing inventory, then whenever this activity occurs a measurement system must exist which will immediately alert the company to undesirable activity. Detailed information on measurement systems can be found in

- *The Haystack Syndrome* (1990)
- *The Theory of Constraints: Applications in Quality and Manu-facturing*
- *Re-Engineering Performance Measurements*

IV.   DEVELOPING THE IMPLEMENTATION PLAN

The implementation of a TOC-based system is very much different than for an MRP system. With TOC, there are the added problems associated with understanding and executing the new decision process as well as eliminating invalid policies. However, invalid policies do not announce themselves. They only appear as undesirable effects. After everyone has been educated, the integration process is complete, but the schedule created there is no guarantee that the implementation will be a success. As an example, what would be the impact if for some reason the constraint kept moving from resource to resource even though everyone was trying their best to follow the schedule? The TOC thinking process is used to ensure that invalid policies can be found and solutions devised to eliminate them. Thus, in developing the implementation plan there are two issues:

1.   Those activities designed to implement system
2.   Those activities necessary to eliminate invalid policies that serve to block a successful implementation

The implementation plan is a plan of action which defines what is to be done, who is to do it, and when it is to take place. Those activities common to every plan for implementing the TOC-based system are

- Begin the educational process.
- Rewrite the corporate policy book.
- Complete the integration to the standard manufacturing system.
- Establish the test net.
- Test the logic.
- Model the plant.
- Determine and execute the plant layout.
- Write the procedures.
- Develop the conference room pilot scenario.
- Execute the conference room pilot.

- Develop and execute the education and training sessions.
- Determine the buffer sizes.
- Determine overtime and setup savings limitations.
- Create the master schedule.
- Develop the initial schedule.
- Execute the schedule.
- Begin the buffer management process.
- Begin TOC decision support.
- Begin TOC thinking process support and help to generate the road map.

Note: Rewriting the corporate policy book includes generating a set of general instructions designed to ensure that managers will use as their guiding principles the methodologies and decision models originating from the theory of constraints.

Each TOC system implementation will also have activities that are very different. For every company, problems will occur that will tend to block a successful implementation. While these activities may have common causes, they will manifest themselves in very different ways. The TOC thinking process is the road map used to guide the implementation process through those activities.

A. The TOC Thinking Process

The objective of the TOC thinking process is to define those actions necessary to improve a company given the current situation and to guide each step to its sometimes not-so-obvious conclusion. It defines

- What to change
- What to change to
- How to accomplish the change

Defined in *It's Not Luck* (1994) and included in the TOC thinking process are five tools:

1. The current reality tree (CRT)—used to find the core cause or causes from undesirable effects.
2. Evaporating clouds (EC) (assumption model)—used to model the assumptions made which block the creation of a breakthrough solution.
3. The future reality tree (FRT)—used to model the changes created after defining breakthrough changes from the evaporating cloud.

4. The prerequisite tree (PT)—used to uncover and solve intermediate obstacles to achieving the goal.
5. The transition tree (TT)—used to define those actions necessary to achieve the goal.

The TOC thinking process is used to guide the implementation of the theory of constraints and to aid in creating breakthrough solutions. This can also be of tremendous benefit to the implementation of the TOC-based system by ensuring that any undesirable effects that manifest themselves during the implementation process are eliminated.

Note: An excellent reference for the TOC thinking process is *The Theory of Constraints and Its Implications for Management Accounting* (1995).

1. The Current Reality Tree

The first step in the execution of the TOC thinking process is to list the undesirable effects (UDEs) and then create the current reality tree. The current reality tree is a diagram built on the cause-and-effect relationships between the undesirable effects and their immediate causes. The objective is to find the core cause. Once found and eliminated, all undesirable effects should disappear.

If the core cause can be found and a solution implemented, then these effects should no longer exist. If they continue, then the core cause(s) was not eliminated. As an example, the following are undesirable effects that may occur during an implementation of the system and can be used to build a current reality tree:

UDE-1: Resource R-3 often receives mismatched parts from R-4.
UDE-2: Sales orders are frequently shipped late for product A.
UDE-3: Resource R-2 often has holes in zone I of the buffer caused by late orders for part/operation A/10.
UDE-4: Part/operations B/30 and D/30 are often expedited from R-4.
UDE-5: There are holes in zone I of the shipping buffer for product A.
UDE-6: R-4 is consistently working overtime to supply parts D/30 and B/30.

Notice that each UDE has been numbered. This is done so that each UDE can be easily identified in the current reality trees to follow. Figure 8 is the product flow diagram in which the previously defined UDEs exist.

Note: The product flow diagram is not part of the thinking process. It has been supplied as a reference point for the discussion of the five tools.

In Fig. 9, UDE-2 is the first undesirable effect to be placed and appears at the head of the arrow. In other words, it is the identifiable effect and can be related directly to the text in block one. Block one and

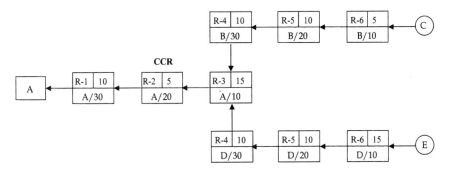

Figure 8 The product flow diagram for product FA.

all blocks numbered in the same fashion are known as "entities." To read the diagram the block at the tail of the arrow is read first. The tail of the arrow is identified with the word *if*, while the head of the arrow is identified with the word *then*. So, the relationship between UDE-2 and entity 1 should read

   if: (entity 1) product A is frequently late arriving from production,
then: (UDE-2) sales orders are frequently shipped late for product A.

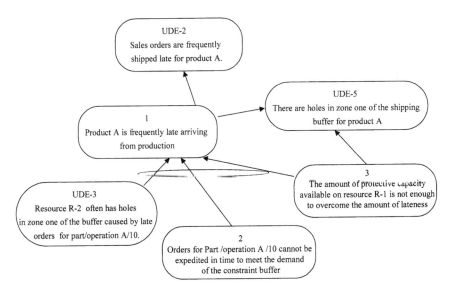

Figure 9 Current reality tree.

Entity 1 is also connected to UDE-5:

    if: (entity 1) product A is frequently late arriving from production,
then: (UDE-5) there are holes in zone I of the shipping buffer for
          product A.

    Note: The process of identifying relationships between UDEs and
entities continues until the core cause is discovered.

    UDE-3 is also connected to entity 1. However, UDE-3 is at the tail of
the arrow. There is now a connection between UDEs 2, 3, and 5 through
entity 1:

    if: (UDE-3) resource R-2 often has holes in zone I of the buffer caused
           by late orders for part/operation A/10,
then: (entity 1) product A is frequently arriving late from production.

Notice that this statement cannot necessarily stand by itself. Just because
there are holes in the buffer for product A does not mean that product A will
be frequently late in arriving from production. There must be supporting
causes. Entity 2 states that orders for part/operation A/10 cannot be
expedited in time to meet the demand of the constraint buffer. (*Note*: The
constraint has been identified in the product flow diagram as R-2.)

    The arrows originating from UDE-3 and entity 2 are joined by an ellipse
along with entity 3. In this case the relationship is read:

      if: (UDE-3) resource R-2 often has holes in zone I of the buffer
            caused by late orders for part/operation A/10,
and if: (entity 2) orders for part/operation A/10 cannot be expedited
            in time to meet the demand of the constraint buffer,
and if: (entity 3) the amount of protective capacity available in resource
            R-1 is not enough to overcome the amount of lateness,
  then: (entity 1) product A is frequently arriving late from production.

In this case, there is enough evidence to point to a definite cause-and-effect
relationship between entity 1, UDE-3, entity 2, and entity 3.

    Once constructed the current reality tree can provide tremendous
insight into the cause-and-effect relationships of activities occurring
throughout the plant. While it is developed from the undesirable effects
and projects downward toward the core causes, the current reality tree
can, and should, be validated by reading from the bottom up. Entity 14 in
Fig. 10 has been identified as the core cause for creating the late orders.

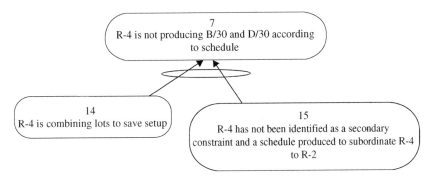

Figure 10   Current reality tree.

When combined with entity 15 there is sufficient cause-and-effect relationship to justify entity 7:

if:	(entity 14) R-4 is combining lots to save setup,
and if:	(entity 15) R-4 has not been identified as a secondary constraint and a schedule produced to subordinate R-4 and R-2,
then:	(entity 7) R-4 is not producing B/30 and D/30 according to schedule.

The issue now turns to understanding why setup savings are being performed for part/operations B/30 and D/30 at resource R-4. Before a successful plan can be adopted a basic understanding of why is important. Short of asking the foreman, a basic analysis should be performed.

(*Note*: Asking the foreman why he is performing setup savings on a nonconstraint resource may solidify his insistence to continue.)

Figure 11 presents one reason that the foreman may be acting as he is. Beginning at the bottom it reads

if:	(entity 18) people will act based on how they are measured,
and if:	(entity 19) R-4 is measured based on productivity,
then:	(entity 17) the actions of people at R-4 will be to maximize the productivity at R-4.

Continuing to the next level, the diagram reads

if:	(entity 17) the actions of people at R-4 will be to maximize the productivity at R-4,
and if:	(entity 20) setup savings is a way to maximize productivity,
then:	(entity 22) R-4 will combine lots to save setup and maximize productivity.

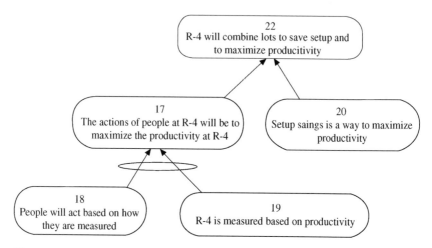

Figure 11   Current reality tree.

So the core cause for the setup savings occurring on a nonconstraint resource may be due to how the foreman is being measured externally (a managerial constraint) or in how he measures himself (a
behavioral constraint).

2.   Creating the Evaporating Cloud

As stated earlier, the evaporating cloud used to model the assumptions made
which block the creation of a breakthrough solution. Figure 12 represents

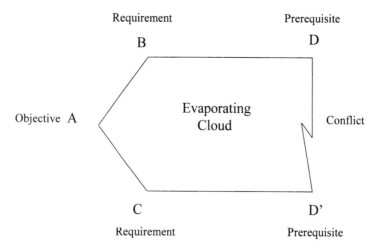

Figure 12   The evaporating cloud.

an evaporating cloud. Notice that the objective of the model is placed on the left at position A. The objective may be to increase return on investment. Positions B and C represent those things that are required in order to meet the objective. Position B might read "increase productivity" and position C might read "exploit the constraint." Position D and D′ represents the conflict. In other words there is an assumption that D and D′ cannot exist together.

The cloud reads as follows:

- In order to have A there must be B.
- In order to have A there must be C.
- A prerequisite to B is that there must be D.
- A prerequisite to C is that there must be D′.
- D and D′ cannot exist together.

To solve a problem involving the evaporating cloud the assumptions made between each of the entries in the cloud are examined for possible flaws. In Fig. 13, in order to say that there is a relationship between A and B there must be certain assumptions made such as the only way to have A is to first have B. If the assumption can be proven false, then the problem evaporates.

   a. *The Foreman's Dilemma: Evaporating the Cloud.* Figure 14 represents the foreman's dilemma. In this case the objective is to increase return on investment. The foreman's cloud reads as follows:

- A-B: In order to increase return on investment, productivity at R-4 must be increased.

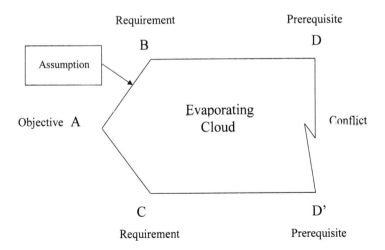

Figure 13   The evaporating cloud—understanding the assumptions.

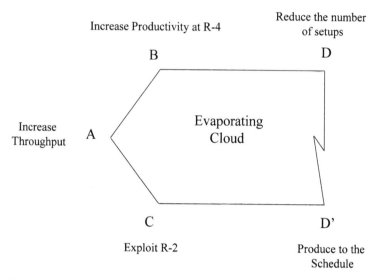

Figure 14   Solving the foreman's dilemma—evaporating the cloud.

- A-C: In order to increase return on investment, R-2 must be exploited.
- B-D: In order to increase productivity at R-4, the number of setups must be reduced.
- C-D′: In order to exploit R-2, R-4 must produce to the schedule.
- D-D′: Producing to the schedule set for R-2 and saving setup are mutually exclusive.

To solve the foreman's dilemma the first step is to list the assumptions being made between the entries. Most of the time there will be more than one assumption made. Our assumptions are

- A-B: There is a direct relationship between the local measurement of productivity at R-4 and the global measurement of return on investment.
- A-C: Exploiting R-2 throughput will increase, thereby causing ROI to increase.
- B-D: Setup savings at R-4 can only have a positive impact.
- C-D′ R-4 must be subordinated to the demand created by R-2.
- D-D′ There is no way that R-4 can follow the schedule for R-2 and perform setup savings at the same time.

A-B makes the assumption that there is a direct relationship between the increased productivity at resource R-4 and the increase in return on investment. This assumption is not necessarily true. Improving at a nonconstraint resource from a global perspective will not cause throughput to increase. B-D also has a problem. As seen later, saving setup on a nonconstraint resource will often subvert the schedule of the constraint, causing ROI to decrease rather than increase. If setup savings is not performed in direct synchronization with the constraint's schedule, then there is a good chance that the constraint schedule will be invalidated by the setup process. This is a key issue in solving the foreman's dilemma.

3.   The Future Reality Tree

As defined earlier, the future reality tree is used to model the changes created after defining the breakthrough solution from the evaporating cloud. In this case, a new measurement system is being provided to the foreman (see Figs. 15 to 17). The future reality tree is built starting with what is called an *injection* and is read in the same manner as the current reality tree. An injection is the change designed to solve the undesirable effects. In this case, the injection is to measure foreman performance based on throughput dollar days. The future reality tree reads as follows:

if: (injection 1) the foreman's performance is measured based on throughput dollar days accumulated,

then: (entity 26) pressure will increase on the foreman to deliver parts according to the schedule for R-2.

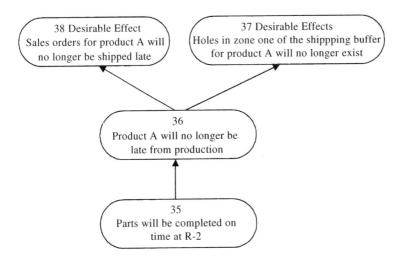

Figure 15   Generating the future reality tree.

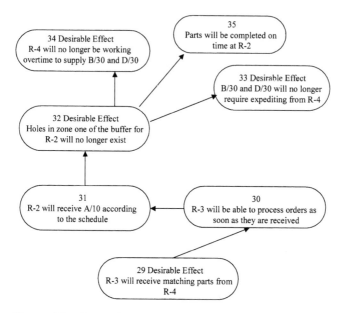

Figure 16   Generating the future reality tree.

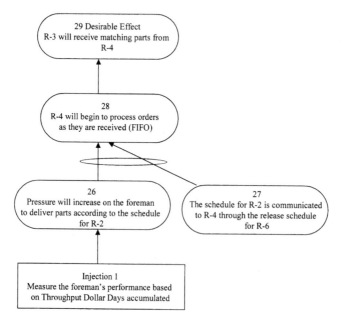

Figure 17   Generating the future reality tree.

And

      if: (entity 26) pressure increases on the foreman to deliver parts according to the schedule for R-2,

and if: (entity 27) the schedule for R-2 is communicated to R-4 through the release schedule for R-6,

  then: (entity 28) R-4 will begin to process orders as they are received: first in, first out (FIFO).

Notice that entity 29 is a desirable effect (DE). This means that the entity is the exact opposite of the undesirable effect from the UDE list. In this case, one of the undesirable effects has been eliminated by the change in the way foreman performance is measured.

   if: (entity 28) R-4 begins to process orders FIFO,

then: (DE-29) R-3 will receive matching parts from R-4.

Notice that in the remainder of the current reality tree all of the undesirable effects have been eliminated, including UDE-2 (sales orders are frequently shipped late for product A), and replaced by desirable effects.

    a.   The Foreman's Dilemma: Trimming the Negative Branches.   Somehow the solution does not seem complete. If the measurement system was not the only issue which prevented the foreman from following the schedule, then the devised solution may not replace all the undesirable effects with desirable effects as defined in the future reality tree. At this point all reservations should be surfaced as to the validity of the solution. Figure 18 serves to illustrate the reservation. The following can be inferred:

      if: (injection 1) the foreman's performance is measured based on throughput dollar days accumulated,

and if: (entity 27) the foreman firmly believes that the problem associated with the lateness of orders is due to a lack of adequate capacity at resource R-4,

But

      if: (entity 25) the foreman continues to perform setup savings for B/30 and D/30,

then: (entity 26) the undesirable effects will not go away.

From Fig. 18 it is learned that there is a distinct possibility that the foreman firmly believes that the undesirable effects are caused by the fact that there is not enough capacity at R-4. Merely placing a new measurement system in this environment may only serve to exacerbate the

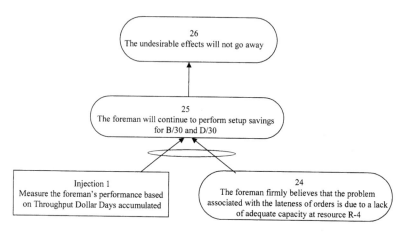

Figure 18   Trimming the negative branches.

situation. Figure 19 defines another possible cause for the setup savings. It reads as follows:

    if: (entity 23) a constraint will sometimes identify itself by the amount of product being expedited or the amount of overtime required,

and if: (UDE-4) part/operations D/30 and B/30 are often expedited from R-4,

and if: (UDE-6) R-4 is consistently working overtime to supply parts D/30 and B/30,

  then: (entity 19) R-4 is perceived to be secondary constraint.

And

    if: (entity 18) people act based on their perception,

and if: (entity 19) R-4 is perceived to be a secondary constraint,

  then: (entity 17) the actions of people at R-4 will be to maximize the productivity at R-4.

Thus

    if: (entity 17) the actions of people at R-4 will be to maximize the productivity at R-4,

and if: (entity 20) setup savings is a way to maximize productivity,

  then: (entity 22) R-4 will combine lots to save setup and to maximize productivity.

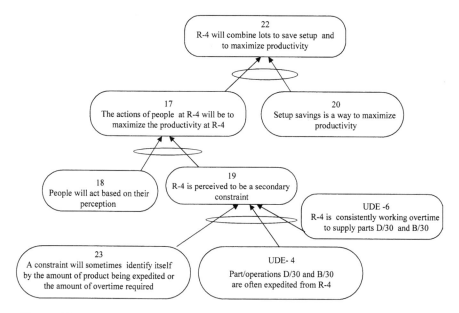

Figure 19 Trimming the negative branches—validating the reservation.

b. Expanding the Future Reality Tree. To solve the new problem the future reality tree must be expanded to include a new entity. The bottom of the future reality tree now reads

if: (injection 1) the foreman's performance is measured based on throughput dollar days accumulated,
then: (entity 26) pressure will increase on the foreman to deliver parts according to the schedule for R-2.

See Fig. 20.

if: (entity 26) pressure increases on the foreman to deliver parts according to the schedule for R-2,
and if: entity 27) the schedule for R-2 is communicated to R-4 through the release schedule for R-6,
and if: (entity 40) the foreman believes that R-4 is not a secondary constraint,

So in order for the problem to be solved, the foreman must believe that R-4 is not a secondary constraint. A new injection must be introduced that will cause the foreman to change this belief. Assuming that the implementation of the system has included education and training, the

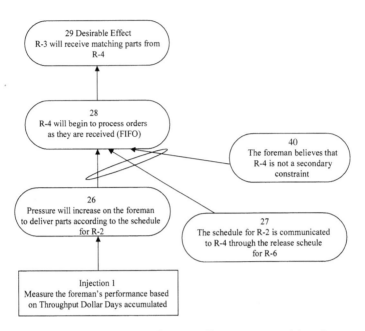

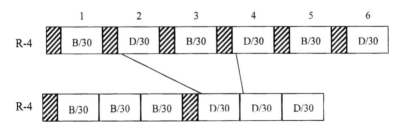

Figure 20 Generating the future reality tree—recognizing the new entity.

foreman should already be aware of the impact of setup savings on a nonconstraint resource. The solution may be in providing a method for the foreman to simulate the current environment to either validate or disprove the belief.

Note: This issue serves to underline the need for the permanent education program so that people can invent their own solutions.

    c. Simulating the Foreman's Problem. Figure 21 defines why the parts arriving at R-3 are mismatched. If setup savings is performed on R-4, some of the orders will be pushed outward in time while others will be pulled in. The result is that R-4, which had been synchronized to the constraint

Figure 21 Simulating the foreman's problem.

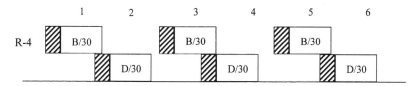

Figure 22 Simulating the foreman's problem—placing orders on the timeline for R-4.

schedule of R-2 by the release schedule for R-6, will no longer be synchronized. According to the schedule in the figure, R-3, will receive large batches of B/30 and then large batches of D/30. The foreman begins working overtime to keep up with material expedited to R-3 and concludes that there is not enough available capacity to remedy the situation. This a normal conclusion and is made every day in the traditional MRP facility.

This is exactly the kind of information that can induce a change in the foreman's activity. Fortunately the TOC-based system is capable of simulating the interface requirements between R-2 and R-4. If the constraint schedule has been created for R-2 and it becomes necessary to determine the impact on R-4, R-4 can be declared a secondary constraint. Orders are then placed on the timeline in synchronization to R-2. If a conflict does not exist between orders on the timeline for R-4 prior to generating a schedule for R-4, the changes are great that there is more than enough capacity to handle the demand generated by R-2. If any conflicts exist, they will surface as two part/operations trying to occupy the same space at the same time on the timeline for R-4 (see Fig. 22). Notice that orders 1 and 2 are trying to occupy the same space on the timeline. However, this does not mean that there is a problem. To understand the issue under this condition a schedule needs to be created for R-4 (see Fig. 23).

If the system places orders so that there are no gaps between orders (additional time), this is an indication that R-4's time is limited. If there are no late orders and setup savings is not indicated on the system, then chances are good that setup savings is not required on the shop floor either. Gaps between orders on the schedule for R-4 are an indication that there is plenty of additional time and that the foreman's assumption that R-4 should be declared a secondary constraint and setup savings performed is false (see

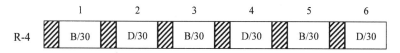

Figure 23 Simulating the foreman's problem—generating the schedule.

Figure 24 Simulating the foreman's problem—gaps between orders.

Fig. 24). Notice that there is a gap between the start of the planning horizon and the start of order 1. Notice also that there are gaps between orders 2 and 3 as well as orders 4 and 5.

    d. Adding the New Injections. There are actually two new injections that are necessary to induce the foreman to stop setup savings at R-4 (see Fig. 25):

> **if:** (injection 2) the foreman is provided with a method to simulate the environment,
>
> **and if:** (entity 43) the simulation indicates that R-4 is not the constraint,
>
> **then:** (entity 40) the foreman will believe that R-4 is not a secondary constraint.

Figure 26 contains injection 3:

> **if:** (injection 3) the foreman is provided with training in the DBR process,
>
> **then:** (entity 41) the foreman will understand the impact of setup savings on a nonconstraint resource.

    In Fig. 27, the bottom of the future reality tree now reads

> **if:** (entity 26) pressure will increase on the foreman to deliver parts according to schedule,
>
> **and if:** (entity 27) the schedule for R-2 is communicated to R-4 through the release schedule for R-6,
>
> **and if:** (entity 40) the foreman believes that R-4 is not the secondary constraint,
>
> **and if:** (entity 41) the foreman understands the impact of setup savings on a nonconstraint resource,

4. The Prerequisite Tree

As stated earlier, the objective of the prerequisite tree (PT) is to uncover and solve intermediate obstacles to achieving the goal. The injections from the future reality tree are used to guide the process. As each injection is presented, obstacles to its realization surface. The intermediate objective in attaining the injection is to overcome the obstacle.

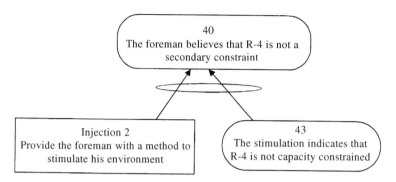

Figure 25 Generating the future reality tree—adding injection 2.

    The injection is usually placed at the top of the diagram in a rectangle, the obstacle is placed to the left contained in a hexagon, while the intermediate objective is contained in a rounded rectangle.

    Figure 28 involves injection 1 (measure the foreman's performance based on throughput dollar days accumulated). Notice that one of the obstacles (obstacle 2) to attaining injection 1 is that the foreman may not understand the benefit of the T$D measurement process. This could easily derail the benefits of using the new measurement system. The other obstacle (obstacle 4) to attaining injection 1 is that the tools necessary to collect and manipulate the T$D data for the new measurement system are not available. To overcome the obstacle the intermediate objective is used. The intermediate objective is the exact opposite of the obstacle. The intermediate objective to overcome obstacle 2 is that the foreman understands the measurement

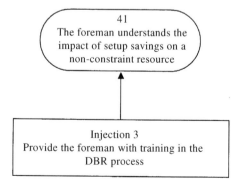

Figure 26 Generating the future reality tree—adding injection 3.

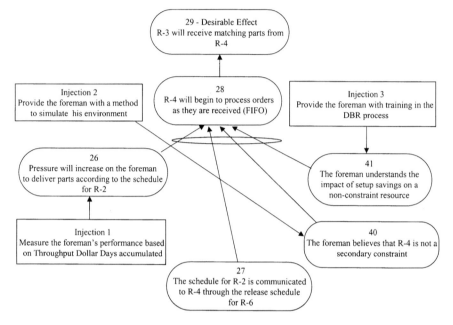

Figure 27   Expending the future reality tree.

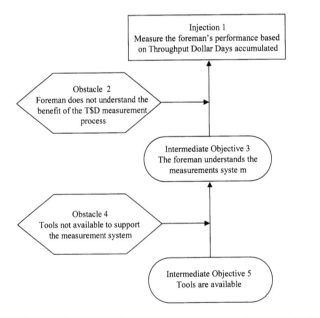

Figure 28   Generating the prerequisite tree for obstacles 2 and 4.

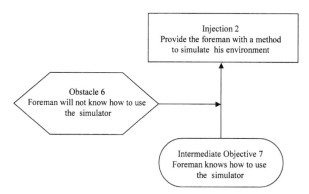

Figure 29 Generating the prerequisite tree for obstacle 6.

system. The opposite of obstacle 4 is that the tools are available. It now becomes clear that in order to obtain injection 1, action must be taken to make sure that the foreman understands the measurement system and that tools be made available to collect data.

In Fig. 29 the obstacle to attaining injection 2 is that the foreman will not know how to use the tool. The intermediate objective is that the foreman does know how to use the simulation tool.

In Fig. 30 the obstacle to attaining injection 3 is that the foreman does not think that the DBR process will work in this environment. The intermediate objective is that the foreman thinks the DBR process will work in this environment.

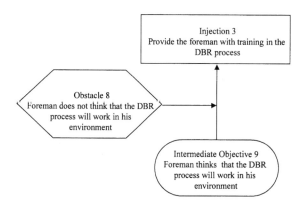

Figure 30 Generating the prerequisite tree for obstacle 8.

5.   The Transition Tree

As defined earlier, the transition tree is used to define those actions necessary to achieve the goal. Like the prerequisite tree, the transition tree begins with the injection (see Fig. 31). The intermediate objectives are listed to the left in the rounded rectangular boxes and the actions designed to attain the intermediate objective are placed in rectangles. The arrows designate which actions apply to the intermediate objective. In order to attain intermediate objective 5, action 1 must take place. In order to have intermediate objective 3, action 2 must take place.

The two actions required to attain the intermediate objectives as well as injection 1 include training the foreman to understand the measurement system and creating the tools to collect and manipulate the T$D data from the shop floor.

In Fig. 32 the action required to insure that the foreman believes that the DBR process will work in this environment (action 3) is to use Socratic method with the educational simulators in the permanent education program and recreate the foreman's environment during education.

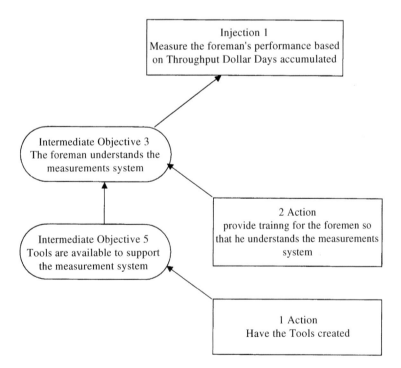

Figure 31   Generating the transition tree for injection 1.

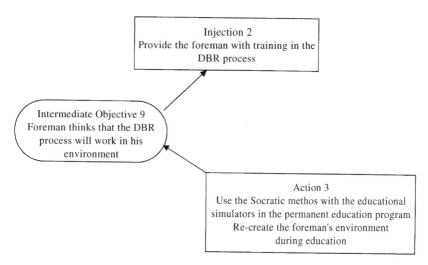

Figure 32 Generating the transition tree for injection 2.

The action required to insure that the foreman knows how to use the simulation tool is (action 4) train the foreman to use the simulator (see Fig. 33).

The TOC thinking process (TP) has been used successfully in companies around the world to create solutions to very complex problems. Using TP to aid in the implementation of the manufacturing information system is a logical extension.

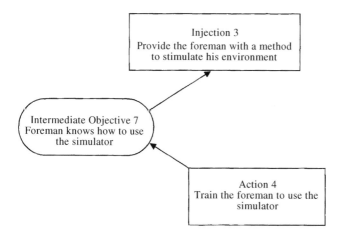

Figure 33 Generating the transition tree for injection 3.

6. Using the System to Develop the Road Map

Since the system is an excellent representation of reality and is devoid of policy constraints, it can be an excellent tool for helping to develop and execute the road map. As an example, the physical environment and its attributes will have a significant impact on solutions provided by the road map. The information system will not only inform the user what the impact of the physical attributes might be, but can also be used to test specific changes suggested in the creation of a breakthrough solution and will also aid in its execution.

The following list suggests ways in which the system can be used in support of the generation of the road map:

- Current reality tree—does excess capacity exist? If so, where?
- Breakthrough solution—testing assumptions.
- Future reality tree—if a specific action is taken, what are the results?
- Prerequisite tree—if a specific goal is identified (e.g., conflicts between resources), what intermediate obstacles will block attaining the goal?

V. SUMMARY

As indicated throughout the text, the TOC-compatible system represents a tremendous paradigm shift from the traditional approach and, as such, will change the way companies operate for years to come. Based on reality, the TOC information system solves a tremendous number of logistical and policy constraints that have hampered the effective use of the computer-based information system since its development more than 30 years ago. Implemented correctly and used to its fullest extent it can be a tremendous asset in support of any manufacturing company.

VI. STUDY QUESTIONS

1. Why is a TOC-based manufacturing system more difficult to implement than a traditional system?
2. Define a successful TOC system implementation.
3. What preparations should be made to prevent failure?
4. Explain the uses and benefits of the permanent education program.
5. What is the TOC thinking process and what advantage does it provide to the implementation process?

6. List and define the five thinking tools.
7. What primary issue will tend to block most implementations and how can it be prevented?
8. Why is stimulation such a great training aid? How should it be used?
9. What key issues are involved in selecting the computer system to run the TOC-based software? Why?
10. What is the benefit provided by the conference room pilot and what issues should be stressed?
11. What is meant by the terms *technical*, *physical*, and *logical implementation*?
12. What is the first step in the physical implementation and what is its objective?
13. What changes occurring on the host manufacturing system result in the need to refresh the net?
14. Develop a complete road map using the TOC thinking process to analyze and solve a problem in your life where your intuition is relatively mature.
15. Define the following terms as they relate to the TOC thinking process:

    - Entity
    - Reservation
    - Undesirable effect
    - Desirable effect
    - Intermediate objective
    - Action
    - Obstacle

14
Implementing an ERP System to Improve Profit

An *enterprise requirements planning* (ERP) implementation methodology designed to focus on improving profits represents a change from the traditional systems implementation approach. In the traditional methodology an attempt is made to improve the company by removing non–value added processes or reducing cycle times. However, in order for return on investment to go up the company must make more money come into the company or less money leave. Reducing non–value added activity does not necessarily accomplish either.

I. CHAPTER OBJECTIVES

- Present a methodology for implementing ERP systems that is focused on improving profit
- Provide the tools necessary for implementation

If an implementation is to be truly successful, assuming that the company wishes to make more money now as well as in the future, it is vitally important to identify where and how actions must be taken to improve. In addition, an acceptance or buy-in of the people who will implement the system is imperative. Steps must be taken to insure user participation through every step of the process. The reason is simple. If users are not involved in creating and implementing the system they will not have faith or ownership in it. There is nothing worse from an implementation's perspective than designing and configuring a system only to have the users reject it during the conference room pilot.

A TOC-based ERP implementation strategy uses a focused approach to identify those things that serve to constrain the amount of money flowing into the company and to increase it by either exploiting or elevating the limitations that exist. Key issues include

- Factory/supply chain synchronization
- Recession proofing
- Product design
- Strategic planning
- Decision-making
- Decision support
- Measurements

A TOC-based ERP implementation strategy also recognizes that cost and operating expense are not the same thing, and by proper focus operating expense can be reduced while maximizing the flow of money entering the company. The overall result is an increase in profit.

II. ORGANIZING THE PROJECT

The objective of organizing the project is to organize resources and activities so that key elements of the implementation are performed in the proper sequence and to insure that proper feedback mechanisms are in place to manage exceptions. Included are the implementation process, the structure, and the documentation mechanisms necessary to keep tract of the who, why, what, when, and where of the project.

A. The Implementation Process

From a global perspective, the major steps of the implementation process, as seen in Fig. 1, are to define the project, analyze requirements, train the project team, develop and test the prototype, train users, and manage go-live activities.

B. The Structure

The structure of the project usually includes

- A project team to perform the actual implementation
- A steering committee to monitor the project team's progress, enhance communication, and make decisions that cannot be resolved at the project team level
- An executive committee to define the overall project goals and to resolve problems that the steering committee cannot

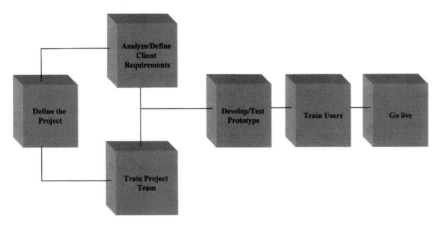

Figure 1   The implementation process.

C.   Documentation

Documentation includes, but is not limited to, the project schedule and any required documentation usually stored in the project binder such as the issues log, technical plans, modification policies, and modification requests.

Note: The organization of a project is a standard within the ERP implementation arena and, by design, leaves much to the interpretation of the implementer and as such can lead in almost any direction.

III.   THE TRADITIONAL APPROACH

In the traditional approach the ERP project is usually define in very broad terms. The assumption is often made that an implementation of ERP will bring about an improvement in profit. So concentration is not placed on this effort, but rather on the applications themselves. After all, justification for a new system was already made during the sales cycle.

When implemented by application, emphasis is placed on the major departments within the company and their desires. Key issues include features and functions such as sales order or work order types, standard versus actual cost, and whether work order issues should be back-flushed or preflushed. Absent is a global perspective that reaches outside the application arena. Goals are established departmentally with the assumption made that if the department can attain its goal then the company should also.

A. The Project Team

The project team's task is to match systems functionality and technology with the project objectives determined by the executive committee, to configure/ implement the system, and to document/report progress. The project team will consist of a project manager, a number of team leads who may also be subject matter experts, and team members. Team members usually consist of consultants who know the software product to be implemented and user representatives.

In the traditional approach project teams are usually aligned by application, such as sales, purchasing, inventory, and production (see Fig. 2). This method is common practice and from the departmental perspective makes sense. The team leader is often the department head, while the members are the users within each department. A functional consultant is usually included on each team to act as a coach to help the users configure the system (preferred), or the consultant will configure the system himself.

Each member of the application team as well as the team leader has a vested interest to insure that his or her department is provided all the features and functionality needed to perform the departmental duties as they are perceived. Any discussions that take place between application teams are usually about what data are needed to accomplish the departmental goal.

Note: No discussion of profit is usually ever conducted within the project team.

The Project Team			
Sales	**Purchasing**	**Inventory**	**Production**
Team Leader	Team Leader	Team Leader	Team Leader
Team Members	Team Members	Team Members	Team Members

Figure 2   Aligning project teams by application.

B. Project Phases

To handle large amounts of data and the enormous change that implementing a new system entails, the implementation process is usually subdivided into smaller pieces of manageable parts. Project phasing is designed to help participants understand the basic direction of the overall project and is used to establish the project schedule. In the traditional approach, the project is again subdivided by application. Phase one may be accounting while phase two and three are relegated to distribution and manufacturing, respectively. Phase content is further subdivided into modules such as purchasing, sales, and inventory for the distribution phase, or general ledger, accounts payable, and accounts receivable for accounting. See Fig 3. (*Note*: Each phase may contain all of the steps included in Fig. 1.)

   When the analysis is performed to map the current business, assigning modules to specific features or to make process improvements, additional attributes are identified such as advanced pricing, shipping, invoicing, and blanket orders. These data are then used to form the implementation schedule. So not only are the project teams and phases aligned by Application, but also the project schedule. The question becomes not what data are needed to support the profit improvement plan, but what data are needed to set up sales order entry. Usually a profit improvement plan does not exist.

C. The Impact of the Traditional Approach

As indicated, the traditional approach lacks and adequate focusing mechanism to insure the implementations achieve their profit goals. Additionally, it takes an inordinate amount of time to determine whether success has been achieved and that what the selection/justification team said would occur to profitability actually happened. In Fig. 3 each phase is scheduled to take 6

Phase	Application	Location/Department	Target Date
I	Accounting	G/L, AP, AR	9/01/02
II	Distribution	Sales, Purchasing, Inventory	3/01/03
III	Manufacturing	Engineering, Production, Production Control	9/01/03

Figure 3 Project phases—the traditional approach.

months. This is a normal schedule in some environments. Why should any company wait 18 months before finding out whether they are going to make more money from their investment?

IV. THE THEORY OF CONSTRAINTS APPROACH

The objective of a TOC-based ERP implementation is to establish an approach that will drastically reduce the amount of time required before payback occurs. To accomplish this goal, focus is shifted away from applications-based implementations, as described earlier, toward a more profit-based implementation. Key issues include how projects are defined relative to their objectives as well as the structure, timing, and content of each phase.

A. Project Definition

For any implementation to be successful it must be defined in a way that is in line with the goal of the corporation. Assuming that the goal is defined as "making more money now as well as in the future," the criteria by which the project is deemed to be a success must rely on profit attainment at the end of the project.

B. Establishing the Objectives

Understanding the project objectives is critically important to determining its overall direction. The question which must be answered from the outset is whether the objective of the project is to increase profit, improve processes, or simply implement the software as is without consideration to changing the way the company operates.

Depending on where in the implementation cycle a company might be, the key decision-maker may only want the system up and running as fast as possible without any changes. Changes that occur to attain an objective of profit improvement may be relegated to phase two of the overall project. The person making a decision to postpone profit improvement objectives is usually the information systems or project manager. Not that they are against making a profit, but because the objectives and due dates have already been set and their primary objectives are to be on time and on budget. The time to set objectives is early in the project and with the involvement of upper management.

Note: If the key decision-maker is still at the top of the company (the CEO or CFO), then profit plays a more important role in determining the overall objective of the project.

Note also: Those project objectives which focus strictly on process improvement as a way to increase profit are usually doomed from the outset to increase operating expense without having a corresponding increase in throughput.

From a management perspective, the project must be defined well enough so that an indication of success or failure can be reached. As an example, the goal may be initially stated as a 4-month break-even. In other words when comparing the amount of increased operating expense due to the cost of the hardware, software, and consulting expenditures the increase in profit will equal the amount paid in 4 months. To accomplish this goal the company may find that a 20% increase in return on investment (entity in Fig. 4) will be required. Since return on investment is equal to throughput minus operating expense divided by inventory, this figure can be easily calculated.

To accomplish entity 4 the company has decided that it will need an increase in throughput of 25% (entity 1), a reduction in inventory of 40% (entity 2), and a reduction in operating expenses of 10% (entity 3).

Note: Expenses associated with the implementation have been isolated from operating expense so that impact on operations can be verified more effectively.

How does a company arrive at these figures without pulling them from thin air? In some cases it is possible to approximate the amount of benefit to be gained from historical evidence. In other words, if inventory reductions of 40% can be produced through the implementation of factory synchronization

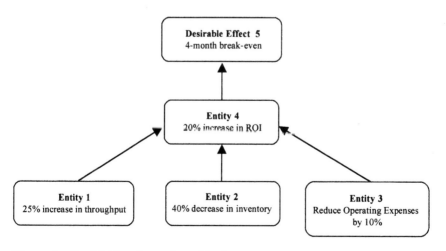

Figure 4   Establishing the project objectives.

(DBR) in environments where MRP is being used in certain industries, then it is reasonable to assume that an implementation in a similar environment would produce a similar result.

In other cases, such as changing the product mix or selling into an additional market segment, the impact may be calculated. It is reasonable to assume that through empirical means a given market can absorb a specific amount of a given product at a given price. Knowing this information allows the implementer to determine the impact to return on investment based on the expected increase in throughput.

In still other circumstances exact calculations may be difficult to compute. Understanding the impact to inventory and operating expense based on a certain decision may best be done through simulation. If in order to attain factory synchronization or to release a new product the company decides it needs to purchase a new machine, it would be advantageous to add the new machine to an existing simulation of the production environment. Of course, simulation may not be available until certain portions of the system have been implemented.

Once the objectives for the project have been set a more reader friendly version should be stored in the project folder (see Fig. 5).

1.   Establishing the Intermediate Objectives

As soon as the project objectives have been established intermediate objectives can then be determined. Meeting intermediate objectives is a prerequisite to meeting the primary objectives of the project. As an example, to

Project Objectives	
1	Increase throughput by 25%
2	Reduce inventory by 40%
3	Reduce operating expenses by 10%

Figure 5   Establishing project objectives.

Intermediate Objectives		
Int. Objective	Obstacle	Action
The factory is synchronized	No constraint schedule	Create schedule for the const.
	No release schedule	Create release schedule
	No buffer management sys.	Create buffer management sys.

Figure 6   Establishing intermediate objectives.

increase throughput by 25% and reduce inventory by 40%, the company may decide to increase factory synchronization and then sell the additional capacity generated.

After the intermediate objectives have been set any obstacle or obstacles to meeting those objectives must be identified and specific actions taken. In the case above, what may be needed is, among other things, the ability to create a schedule for the constraint, a release schedule for gating operations, and a buffer management system (see Fig. 6). Review Chapter 13 for more information on the TOC thinking process.

2. Establishing Prerequisites and Application Relationships

When the actions necessary to meet the intermediate objectives have been identified a road map can be created identifying those tasks required to accomplish the actions. As an example, in order to create a schedule for the constraint, as seen in Chapter 7, there must be part numbers, sales order numbers, sales order quantities, sales order due dates, part/operations, quantities on band, setup times, and run times. Once these prerequisites have been identified a direct connection can then be made to the required applications within the ERP system (see Fig. 7).

These applications are used in the creation of the task list and priorities for the project plan. In this way the project manager can be sure that there is a direct relationship between the actions and priorities of the ERP implementation team and the goal of the implementation. In the above scenario

Application Requirements	
Prerequisite	**Application**
Part Number	Item Master/Item Branch
Sales Order Number	Sales Order Entry
Sales Order Quantity	Sales Order Entry
Sales Order Due Date	Sales Order Entry
Part/Operations	Bill of Material, Routing
Quantity on Hand	Inventory
Setup Time	Routing
Run Time	Routing

Figure 7   Generating the application requirements.

creating factory synchronization is given the highest priority. The faster the factory becomes synchronized and excess capacity sold, the earlier the project's goal will be attained. In order to focus efforts on the profit objective, features such as advanced pricing, engineering change management, and service order management would be relegated to another phase.

C. Setting Priorities

The traditional method for setting priorities, as seen earlier, has proven to take an inordinate amount of time and effort with little, if any, benefit relative to the goal of the company. It is too widely focused to be effective. For the implementer to attain a goal of increased profit he/she must first narrow efforts to a specific goal and its requirements. Success depends on a more coordinated approach that synchronizes very specific applications and data within the system.

Project phases should represent the actions required to attain those goals. When a specific phase is complete the profit of the company should go up (see Fig. 8).

Rather than looking at the system to plan the implementation, the implementer must first look at the company, determine what improvements are required, and then look at the system to determine how to accomplish the

Phase	Improvement Strategy	Location/Department	Target Date
I	Factory Synchronization	Sales, Planning, Production, Production Control, Manufacturing Engineering	9/1/02
II	Decision-Making/Support	Sales, Production, Inventory, Purchasing, Distribution	12/1/02
III	Recession-Proofing	Sales and Marketing, Production	3/1/03
IV	Strategic Planning	Corporate Office	6/1/03

Figure 8 Setting priorities and establishing project phases.

goal. Setting project phases should be accomplished by determining what improvements are required. The first requirement identified in Fig. 8 is factory synchronization and is assigned phase one. The next step is to give the company the ability to know where excess capacity exists so that it can be sold and at what price. Phase two is dedicated to decision support.

Note: An assumption is made here that in order to continue to improve a correct decision process is needed. A more focused approach may be necessary to identify the exact decision process that requires improving before profit goes up. If this is the case, more phases or subphases will be needed.

Characteristics and benefits of this type of implementation include

- Phases are shorter in duration.
- Profit improvements are gained iteratively and more quickly.
- Improvements are more predictable.
- Phases are action/result oriented rather than module/application oriented.

D. Project Synchronization

The objective of project synchronization is to insure that each phase is completed in the shortest period of time possible while remaining on time and on budget. The traditional method of project scheduling and management fails to accomplish this task effectively. The result is constant rescheduling and budget overruns. The critical chain project management strategy insures that projects stay on schedule by helping to identify and manage what is called the *critical chain*. It also buffers the systems so that key activities are properly protected from the inevitable delays that hinder project synchronization. Additionally, critical chain management strategies

Phase	Sales	Manufacturing	Purchasing	Inventory
Factory Synchronization	Sales Order Entry Shipping ——— ——— ——— ———	Bills of material Routings Work order entry Planning Scheduling ———	P/O Entry Receiving ——— ——— ——— ———	Part Maintenance Issues Receipts ——— ——— ———

Figure 9   Aligning application requirements with the phases.

identify and eliminate behavioral issues that sabotage the implementation (see Chapter 15).

E.   Aligning Applications with the Phases

After the phases have been identified and so that project team members have a better understanding of what is required, the application and data requirements are aligned with the phases (see Fig. 9).

F.   Setting Project Expectations: Creating the Profit Improvement Plan

It is important to relay the objectives of the implementation to the executive committee, steering committee, and project team in an effective and unambiguous way. It is also important to know when success has been attained (see Fig. 10).

Phase	Objective	Throughput Chain ID	Constraint ID	Constraint Owner	Additional O/E Required	Break-Even
1	Factory Synchronization	PC Boards	Logistical	Production	$500,000	9 Mo.
2	Decision Support	PC Boards	Market	Sales	$200,000	5 Mo
3	Recession Proofing					
4	Strategic Planning					

Figure 10   The profit improvement plan.

The profit improvement plan is designed to provide the objective, the identity of the throughput chain, the constraint, the organization that owns the constraint, the investment required, and the expected results.

For example, in phase one it was determined that the constraint for a printed circuit board facility was logistical in nature and that production needed to establish factory synchronization to solve the problem. In doing so, the company would invest $500,000 in software, hardware, and consulting fees and reach the break-even point at the end of 9 months. Phase two would spend an additional $200,000 to implement/modify additional portions of the system that would support the decision process and reach break-even at 5 months.

G. Project Concerns and Limitations

Project limitations refer generally to time limitations such as having a requirement that the accounting implementation must be complete by a certain date. They can also include issues such as resource availability and budgetary considerations. The amount of money available to support the project may limit the amount of outside help that the project manager can expect. Any outside influence that limits how the project should be scheduled is referred to as a limitation and should be reflected in the project schedule.

A concern identifies any threats to a projects success or failure. In Fig. 11 a strong cost accounting presence will threaten the project team's ability to synchronize the factory. The reason is given as a lack of education. However, visibility of a concern is also an issue: if it is not known by all involved it will not be addressed properly. Issues and concerns should be placed in the project folder, communicated to the steering committee and to the project team, and addressed in the project plan.

Concern/Limitation	Reason
The project must be finished by 9/30.	New fiscal year begins 10/1.
Team members are 50% available.	They will continue in current jobs.
EDI sales order entry available by 8/1	Customer requirement
Strong cost accounting presence	Lack of education

Figure 11 Concerns and limitations.

H. The Project Team

Unlike the traditional approach, project teams should be aligned by the project phase and the goals of the phase. Instead of having sales, purchasing, or manufacturing teams, the teams should be aligned with the objectives to be attained (see Fig. 12).

The factory synchronization team includes representatives from those organizations that are responsible for the data required to implement phase one. The decision support team includes those required to accomplish phase two and so on. Instead of having six or eight separate teams focusing on departmental requirements, there is one team focusing on the goal of each phase.

I. The Executive Committee

To establish the project goal and objectives it is essential that an executive committee format be used. Depending on the size of the company, the executive committee would include, but not necessarily be limited to, the president and vice-presidents. For the project, the main focus of the executive committee is to determine the project objectives, establish how they are to be reached, and solve any problems that cannot be handled at the steering committee level (see Fig. 13).

The advantage to using the executive committee format is that objections can be surfaced and decisions made quickly at the highest levels within the company, and it ensures input from all departments. In the TOC vernacular the advantage is to ensure that the five steps are implemented

The Project Team			
Factory Synchronization	**Decision Support**	**Recession-Proofing**	**Strategic Planning**
Team Leader	**Team Leader**	**Team Leader**	**Team Leader**
Team Members	**Team Members**	**Team Members**	**Team Members**
Sales and Marketing	Sales and Marketing	Sales and Marketing	Sales and Marketing
Planning	Engineering	Engineering	Engineering
Production Control	Purchasing	Production	Production
Engineering	Quality Assurance		
Inventory	Maintenance		
Purchasing	Production		

Figure 12 Establishing the project team.

Executive Committee		
Member	**Title**	**Role**
	President	Chairman
	VP Sales	
	VP Manufacturing	Executive Sponsor
	VP Distribution	
	VP Engineering	
	VP Information systems	

Figure 13   Establishing the executive committee.

effectively without encountering blocking actions from one or more departments involved in the implementation. As an example, the CFO may want the
project to begin with financials. This could result in delaying the attainment of
factory synchronization. The controller could insist on an implementation of
cost accounting. Cost accounting has the distinct ability to eliminate synchronization. In an executive committee format these challenges can be met
and resolutions derived quickly to ensure success.

J.   The Steering Committee

The steering committee consists of, but is not limited to, representatives from
each of the departments affected by implementing the system, the project

Steering Committee		
Member	**Title**	**Role**
	Executive Sponsor	
	Chairperson	
	Project Manager	
	Departmental Managers	

Figure 14   Establishing the steering committee.

manager, a representative from MIS, and a representative from the executive committee. It will also include a chairperson to manage the activities of the committee (see Fig. 14).

The steering committee is used to carry out the wishes of the executive committee in attaining its objectives, defining the makeup of the imple-mentation teams, and overseeing the activities of the project team. The steering committee meets on a monthly basis to discuss the activities of the implementation team. One of the major benefits is to promote com-munication between departments and to prevent managerial constraints from occurring.

K. Training the Project Team

For project team members to be effective they will obviously need training, from a functional as well as a technical perspective. Unfortunately, the training available to most project team members is application specific. Additionally, vendor training is designed to give the project team an under-standing of what is contained in certain modules, not to get them to the point where they can implement the software without help. And it is very difficult to pass all the information required to understand a module like sales order management in a 5-day class.

To accomplish training for a specific phase will more than likely take the form of custom training. This training should be based on the application requirements of the particular phase being addressed. The trainer/course developer must remember that the objective of the class is not to teach students how to perform sales order or purchase order entry, but to give them enough information to accomplish the goal of a specific phase. As an example, phase one would include classes on factory synchronization and understand-ing the dependent variable environment and behavior/managerial constraints associated with scheduling a factory. This would be followed by training that takes the concepts identified in the educational portions of the class and applies them to the system being implemented.

L. Modification Policy

Because of the nature of the traditional systems currently on the market, a conversion to a TOC-based system will surely require some modification. While most requirements can be addressed through report writing or the generation of inquiry screens, any changes to the system should be justified based on the goal of the phase. A policy statement regarding software modifications should be written by the project manager and approved by the steering committee.

A typical policy statement may indicate that no modifications are to be made unless specifically authorized by the project manager or steering committee. Specific policies should include

- Position Statement: if or when modifications should be made
- Justification Policy: how a modification is to be justified
- Approval Policy: who will approve modifications
- Development Policy: who will develop/test and to what standard
- Maintenance Policy: how modifications will be maintained and by whom

Whenever modifications are requested they should be justified based on the impact to the phase.

Note: An assumption is made here that the phase was throughput justified prior to being established.

V. SUMMARY

That the implementation of an ERP/supply chain system will increase the profitability of a company is a major assumption made by most systems selection teams. Unfortunately, the team's primary emphasis during the selection process is usually placed on functionality, not on profit. This emphasis and its related assumption does not stop at selection but continues on throughout the implementation process and results in profits going down rather than up.

A profit-based ERP implementation strategy will make a tremendous difference from a financial perspective in whether or not the project is deemed a success or failure. By establishing profit objectives for each phase and then aligning the implementation strategy accordingly, profit improvements become more predictable and are attained more quickly. Ignoring these issues will, more than likely, reduce profits by increasing operating expense without a corresponding increase in throughput. Additionally, the objectives of the implementation must be set at an early stage and have buy-in from managers and users alike. The steps to a successful ERP implementation include

- Establish the phase objectives.
- Establish the intermediate objectives, obstacles, and actions.
- Define the prerequisites and application relationships.
- Generate and document the profit improvement plan.
- Develop the phase and critical chain schedule.
- Implement and monitor the phase through go-live.
- Determine whether phase objectives have been met.

Following these steps will drastically change the impact of an ERP implementation and lead companies to increase profitability more quickly. Additionally, profit improvements will continue well after the implementation is completed.

VI.   STUDY QUESTIONS

1. Why do most ERP implementations fail to increase profits?
2. What are the steps necessary to produce a project schedule that is aligned with the goal of a company?
3. What changes must take place in the traditional implementation process to insure that project objectives are met?
4. What are the greatest challenges to a successful ERP implementation?
5. What are the benefits of the methodology suggested in this chapter?
6. What is the purpose of the profit improvement plan and how is it established?
7. What are the steps to a profit-based ERP implementation plan?

15
Critical Chain Project Management

In the past few years and with the efforts of many managers (such as Rob Newbold of Prochain Solutions, Inc.) and their staffs, project management strategies and technologies have made some interesting changes for the better. This chapter describes why change is necessary and defines the methodologies and structure required to better help project managers and companies meet their goals.

I. CHAPTER OBJECTIVES

- To address the shortcomings of current project management strategies
- To present new strategies designed to greatly improve on time and budget measurements

II. WHY PROJECTS FAIL

The reasons for project failure are manifold; some examples are

- Failure to adequately define the scope and goals of the project during analysis and design
- Failure to properly staff the project with qualified people
- Failure to stay within the scope and goals defined by the project (scope creep)
- Lack of user participation

The list is endless. However, much can be said about issues generally not covered in the traditional literature that point toward different causes:

- Improper use of time buffers
- Multitasking
- Resource activation rules
- Improper focusing mechanisms
- Resource contention

A great many projects fail to meet the desired due date because of a lack of understanding of the cause-and-effect relationships that govern the project environment. This lack of understanding easily results in projects being poorly synchronized or improperly focused.

With projects, as in the factory, synchronization is of seminal importance. Any behavior degrading the impact and the effectiveness of projects synchronization will also have a direct impact on the ability of the company to meet its goal. Improper focusing mechanisms make it very difficult, if not impossible, to bring the project back on schedule.

III. DEFINING THE CRITICAL CHAIN

It is not enough to say that projects must be synchronized. The issue becomes how? In order to determine this it becomes important to understand what actions prevent synchronization. There are very common threads that connect the issues involving the synchronization of the factory and the synchronization of the project. The first is that, like scheduling the factory, scheduling the project must account for the fact that resources interface with each other and that they may do so in very complex ways. Each resource will have an ability to produce at a certain level of output and will fluctuate in ability and result. This being said, it is assumed that a certain amount of protection will be required to insure that when a problem occurs it can be dealt with adequately.

The second common thread is that, like the factory, because there is a chain of events there must be a constraint somewhere within the chain. Otherwise the company would be making an infinite amount of money. The question is will the constraint in a project be the same as in a factory? Will the definition of the constraint effectively transfer from a company that manufactures products to one that manages projects?

To understand this will require revisiting the definition of the constraint. The primary constraint is anything which limits the system with regard to achieving its goal.

In a for-profit company that specializes in projects, the goal is still the same: to make more money now as well as in the future. How do companies that manage projects insure that more money enters the company?

For many companies, money is received when a project is completed. For others money enters the company at interim pay points. It stands to reason that the faster a project can be sold and completed or a pay point attained, the faster money will enter the company. So the constraint can be defined as those tasks that limit the further reduction of the lifespan of the project. With respect to scheduling and completing the project, in traditional terms, this is referred to as the critical path. In manufacturing terms this is referred to as the cumulative lead time.

In projects, as in manufacturing, there may also be resource limitations. It is the impact of resource limitations and the critical path which are defined by the Theory of Constraints (TOC) as the critical chain.

IV. UNDERSTANDING RESOURCE CONTENTION

One of the interesting characteristics about projects is that, unlike manufacturing environments, they have no defined routing. While a basic task flow is known, the details are usually unknown or are at least nonstandardized at the beginning of the project. This is why one of the distinctions between factory synchronization and project synchronization is that not only must the tasks be defined, but the critical ones that determine the length of the project must also be identified.

The flow of a typical software project can be defined in high-level terms as analyze, design, construct, test, and install. When the detailed tasks have been identified this flow will repeat itself many times throughout the project. However, each phase within the flow may use the same resources to accomplish the individual tasks. As an example, the same programmers

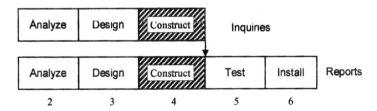

Figure 1 Resource contention.

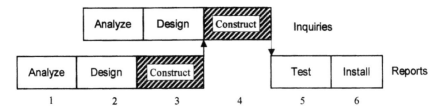

Figure 2 Resource contention.

may be used to construct the inquiry screens for a particular application as well as the reports provided to management.

In Fig. 1, Because the construct tasks are scheduled at the same time (period 4) and because the same resources are used there is a resource contention. The same resources cannot perform both tasks during the same time period. This does not mean that there is a constraint in programming. It only means that during period 4 programming must make a decision to construct either the reports or the inquiries, but not both.

In order to solve the resource contention problem, one of the construction tasks must be moved into the future or the past (see Fig. 2). Report construction is moved into an earlier period so that the integrity of the due date of the project is not compromised. However, the length of the project is increased to six periods from five. This means that the critical path has also been increased to include resolution of resource contention issues. The critical path now becomes the critical chain (see Fig. 3).

A. Multitasking

Most companies tend to solve the resource contention problem by performing multitasking. Multitasking refers to the attempt to perform more than one task at the same time. Unfortunately, this only serves to push out the delivery

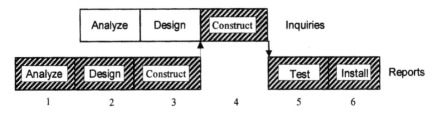

Figure 3 The critical chain.

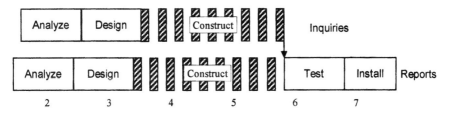

Figure 4 Multitasking.

date, not necessarily because the time required to perform the construct task is squeezed into a space within the schedule but because multitasking increases the amount of time to perform tasks simultaneously. Figure 4 represents the impact of processing both construct times simultaneously. Notice that the completion time has been extended to the end of period seven.

However, what is not accounted for is the additional time that multitasking will add. Like setup time on a machine in manufacturing, multitasking adds setup time between tasks. Even if all you do is change your mind multitasking can add up to 100% additional time to the project for the tasks performed.

In Fig. 5 it's easy to see the impact of multitasking. However, in reality this additional time is hidden and only becomes noticed as project dates slip.

B.   Time Buffers

Time buffers refer to the time used to insure that those things which can and will go wrong do not have a negative impact on the attainment of the project schedule.

When asked for time estimates, most people have a natural tendency to extend the amount of time they have to perform their specific tasks so that they won't be late. The curve in Fig. 6 represents two different answers to the question how long will it take? Time moves from left to right. The 90% figure

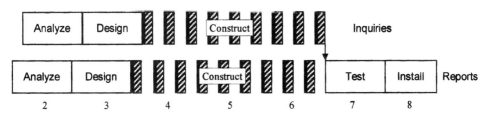

Figure 5 Multitasking.

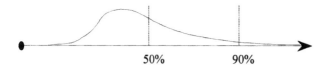

Figure 6   Estimating task times.

indicates an answer that insures that 90% of the time the task or group of tasks will be finished on time. The 50% figure indicates an answer that insures that 50% of the time the tasks will be finished on time.

Most people when asked how long it will take to complete a task will give an answer that will insure a 90% on-time attainment, primarily because they do not want to be late with their portion of the project. However, since the tendency is for everyone in the project to give exaggerated times the result is a schedule that buffers every operation. When every operation is buffered, it becomes impossible to judge how much buffer is available and how much has been used in reference to the success of the entire project. It is also impossible to determine what the priorities should be.

Figures 7 and 8 represent the issue in graphical terms. Production times and buffer times have been split so that it is easy to see the problem. Each task estimate is given in the 90% probability range. Since most people will wait until the last minute to start, task buffers have been placed to the left and production times have been placed to the right. Time flows from left to right.

Because there is no distinction between the start and stop times of the buffer and production time, specific times for actually starting each task are not given in a traditional project schedule. And since most people will be multitasking during the buffer time it is very difficult to determine actual priorities between the task to be performed and the other tasks.

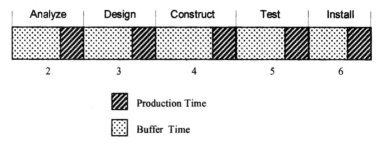

Figure 7   Production versus buffer time.

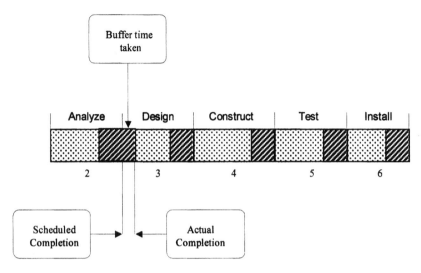

Figure 8 Buffer consumption.

Note: It is also fair to consider that in attempting to perform all the different tasks at the same time additional time will be added that was never considered in the original estimates. However, that is not the primary issue of this section.

Because it is hard to determine how much buffer there is in the project, it is impossible to determine the impact of using a given portion. It would be like asking the question how long is a piece of string? The answer to the question is somewhere between zero and infinity. Unless you know how much string there is and what it's to be used for it is impossible to determine what the impact is of removing one meter.

Figure 8 shows the impact on the buffer by having the analysis portion of the project be completed late. In this case, the buffer taken from the remaining project tasks is reflected in the amount of lateness of the analysis phase (actual completion minus scheduled completion).

The analysis phase is scheduled to be completed by the end of period 2. However, approximately 25% of the buffer for the design task has been used to complete the analysis task.

One of the primary issues involved in managing projects is to know the point that project delivery becomes threatened. In order to do this the amount of buffer must be identified and the relationship of lateness to buffer consumption must be known. This means that buffer time must be isolated from production time and a method devised to show buffer consumption.

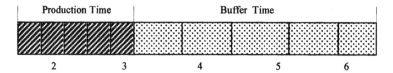

Figure 9 Isolating the buffer.

In Fig. 9 production time and buffer time have been isolated from each other. Start and stop times for each task are given without consideration for buffer time. Buffer time appears between the last task of the project and the due date of the project. Since there is a start and stop time for each task and since the buffer is now a known entity, the impact of buffer consumption can be determined.

V. THE BUFFERING SYSTEM

Critical chain project management provides three different types of buffers: project buffers to protect the entire project, feeding buffers to protect the critical chain, and resource buffers to warn resources used in the critical chain that they have a task to perform.

A. The Project Buffer

Figure 10 shows the project buffer, leading from the end of the critical chain to the due date of the project. The critical chain has been determined without the use of time buffers. The task times include only those times required to complete each task without problems. The project buffer must be large enough to insure that the accumulation of "Murphy" throughout the project will not result in a delay in the due date of the project.

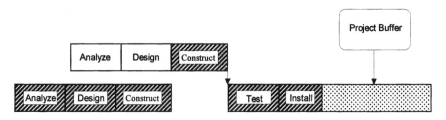

Figure 10 The project buffer.

B. The Feeding Buffer

It is important to insure that those portions of the project feeding the critical chain are themselves buffered so that there are no delays in starting the critical chain tasks. Any delay here will result in threatening the on-time completion of the critical chain.

The analysis and design phases in the upper leg of Fig. 11 do not fall within the critical chain. However, they do feed the upper construct task and that task is on the critical chain. It is very important to the on-time completion of the project that the upper design phase finishes on or before the time that the upper construct task is to begin. To insure this will happen the feeding buffer must be large enough to offset for those things that can and will go wrong in the upper analyze and design phases.

C. The Resource Buffer

As is often the case, most resources on the critical chain will be busy accomplishing other tasks prior to starting each job on the critical chain. The resource buffer is designed to warn the resource that a task is to be performed on the critical chain so that the resource can be prepared to start when the previous task is completed.

The R flags in Fig. 12 designate the resource buffers. Notice that a resource buffer precedes every task on the critical chain. The resource buffer should be large enough so that resources designated to begin the task can make all preparations necessary prior to the completion of the preceding task. Notification is given as a report or e-mail message to the critical chain resource at a certain amount of time before the task is scheduled to start.

Having been given notification that his task is to start soon, the critical chain resource can communicate with the previous task resource to determine

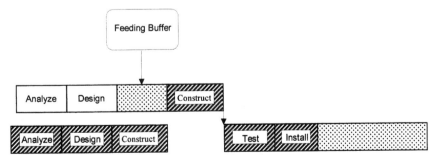

Figure 11   The feeding buffer.

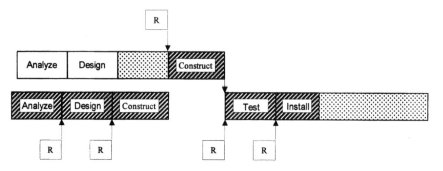

Figure 12   The resource buffers.

what information or tools might be needed to accomplish his task or to insure that the previous task will complete on time.

Note: Resource buffers are no longer recommended or required by current TOC standards. Practitioners have found that simple schedule reports are adequate.

D.   Consuming the Buffer

In Fig. 13, the consumption line indicates that the first task has been completed along with one-third of task 2. However, time zero is toward the end of period 2 indicating that task 2 should be two-thirds complete. Because there is a gap between time zero and the consumption line, some portion of the buffer has been consumed.

The buffer consumption line reflects the impact to the project of task 1 being completed late. Notice that the impact to the overall buffer is minimal.

Figure 14 reflects the impact to the buffer when a larger portion is taken. It's easy to see that approximately 70% of the buffer is taken. It's also easy to see that this project is in trouble. It was supposed to have one-third of

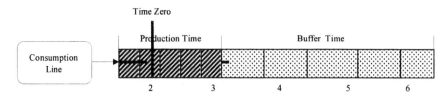

Figure 13   Consuming the buffer.

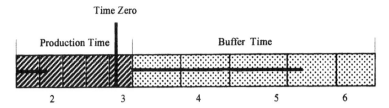

Figure 14   Consuming the buffer.

task 5 complete, but has only completed the first task and one-third of the second task.

E.   Understanding Measurements

One of the key objectives of the buffering system is to show the impact of "Murphy" on the project, but at what point is the probability of finishing the project on time compromised? The traditional method is to look at the amount of time left to complete the project with the amount of time available. However, since the buffer is hidden it is impossible to understand how much of the time left is production time or buffer time. It becomes very difficult to understand the probability that a project will complete on time.

Critical chain management provides measurements relating to how much buffer is left compared to the amount of time left to complete the project. Key indicators are the percentage of buffer remaining compared to the amount of time left in the buffered tasks.

As seen in Fig. 15 the analysis phase of the project has not been completed while time zero is a third of the way through design. A feeding buffer has been set to protect the construct task. However, only 20% of the feeding buffer remains, while 80% of the task time preceding the buffer must

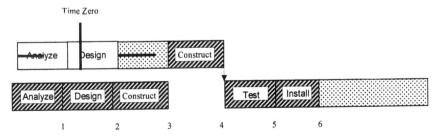

Figure 15   Consuming the feeding buffer.

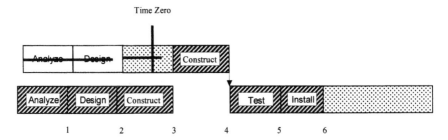

Figure 16   Consuming the feeding buffer.

still be completed. It is highly likely that the design task will finish late, thereby threatening the critical chain.

In Fig. 16 the analysis phase has been completed and the design phase is 80% complete. There is a good chance the project will complete on time even though 80% of the buffer has been used.

F.   Resource Activation Rules

As in manufacturing, each task within a project has a certain probability that it will be completed either early or late. If the buffer has been removed and placed at the end of the project, as discussed earlier, this means that 50% of the time tasks within the project will be early and 50% of the time tasks will be late, (see Fig. 17).

Buffers are placed in the project to offset for the occurrence of "Murphy" (those things that will cause delays and are not foreseen). If tasks are never allowed to be early, they will begin accumulating the probability of lateness.

The best way to schedule tasks is to start gating tasks, those tasks having no preceding tasks, on time and then finish each task in first in, first out (FIFO) fashion quickly as possible until the project is complete. This ensures that project and feeding buffers are consumed only for reasons that

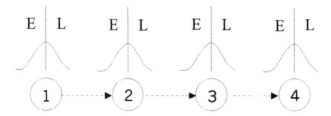

Figure 17   Resource activation rules—probability of lateness for four tasks.

are not controllable. Activation rules are controllable and therefore should not contribute to buffer consumption.

VI.   THE PROCESS OF CONTINUOUS IMPROVEMENT

The process of continuous improvement for projects, much like that found in manufacturing, includes the elimination of invalid policy constraints. As applied to the critical chain, the objective is to eliminate the invalid policy and shorten the critical chain. Once shortened, money enters the company faster.

The following list is taken from *Project Management in the Fast Lane: Applying the Theory of Constraints* (1988) by Rob Newbold and defines the process of continuous improvement for project scheduling:

- Level the load.
- Identify the critical chain.
- Challenge the assumptions of the critical chain.
- Change the project to reflect the changes in assumptions.

The first step is to level the load so that resource contentions can be eliminated. Once eliminated, the next step to identify the critical path is performed and the critical chain is identified. After identification takes place the critical chain is reviewed for any invalid assumptions. If any assumption can be broken, then the critical chain can be shortened.

Figure 18 is a project schedule for building a house created with Microsoft® Project2000® and the aid of ProChain® Pipeline®. ProChain Pipeline is a critical chain compliant software add-on developed by ProChain Solutions, Inc.

Included in the schedule are one project buffer (PB), two feeding buffers (FB), and eight resources buffers (RB). The project buffer, task number 28, has been placed at the end of the project just after install fixtures and appliances. It is used to buffer the entire project and stretches 9 days from 6/14/02 to 6/27/02. Task number 10 is a feeding buffer and is used to insure that the painter finishes task number 9, painting the exterior of the house, before the start time for painting the interior, task number 21.

Figure 19 represents an isolation of those tasks that appear on the critical chain. Notice that task number 21, paint interior, is on the critical chain.

A.   Progress Reporting

Traditional project management systems provide a method of reporting the progress of project tasks by reporting the percent completed of each task. However, in systems development, many developers have the tendency to

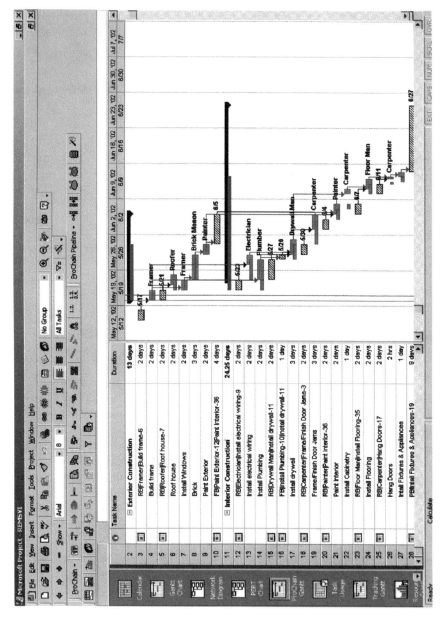

Figure 18 Example of a project schedule. (Microsoft® and Microsoft Projects2000® are registered trademarks of the Microsoft Corporation. All rights reserved. Screen shots reprinted by permission of the Microsoft Corporation.)

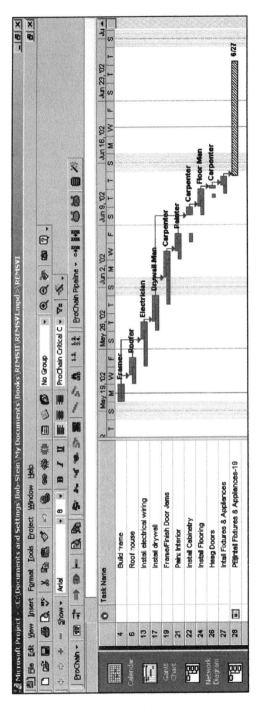

Figure 19 Identifying the critical chain.

work on those tasks that are easy to complete while postponing the more difficult to a later time.

In percentage terms 90% of the tasks may be complete for a specific project deliverable, but the majority of the time left to complete may be in the more difficult tasks. As the project begins to slip, features are eliminated because there is not enough time left to deliver. In software projects the result is often less functionality delivered to the market. This tendency exists because of the way success is measured. To the developer and management it sounds better to say that 90% of the tasks were complete before time ran out and we had to deliver the product to the market. To the market the software company appears to meet its commitments.

Additionally, and probably most importantly, percent complete has been shown to be a meaningless measurement. The assumption made is that percent complete can tell how much time is left to complete, but this may not be the case at all. As an example, if a car that started at point A was originally traveling at 60 MPH but reduced its speed to 30 MPH at the half-way point due to mechanical problems what would be the advantage of saying that they had completed 50% of the journey? If the original distance was 100 miles, the situation is better understood if it is said that there are 50 miles left to go and we have mechanical problems.

Critical chain project management recommends reporting the amount of time left to complete each task rather than reporting the percentage of tasks complete.

B.   High-Level Versus Detailed Schedules

One of the biggest problems facing project managers is to understand what is important to keep the schedule on track. In order to accomplish this the project manager must be able to separate the important activity and accomplishments from the less important. This task is made more difficult because project managers tend to schedule projects in minute detail. However, whenever the number of tasks scheduled within the project reaches approximately 200 the ability to remain focused declines. The project manager quickly loses the ability to maintain a high-level concept of how well the project is performing relative to the goal.

Project plans should be prepared at a high enough level to include major milestones or tasks and exclude the scheduling of detailed subtasks. Alternatively, subtasks can be managed one of three ways:

1.   The use of checklists that identify subtasks needing to be performed within a given task on the high-level schedule.

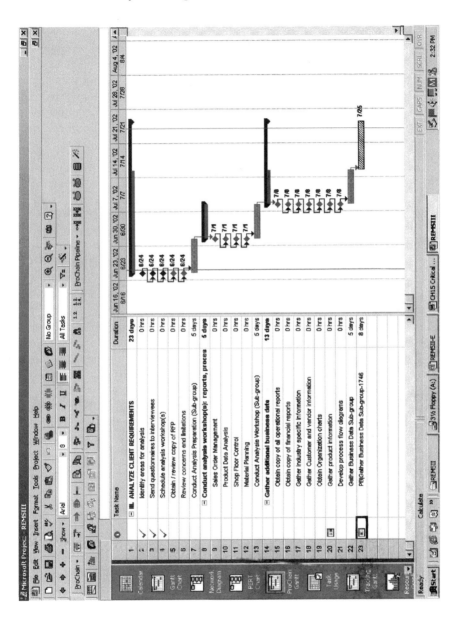

Figure 20 Using a checklist to reduce complexity.

2. The use of subschedules that provide schedules for subtasks within specific milestones of the high-level schedule. The subschedule is used to set priorities while the high-level schedule is used to determine the health of the project.

3. A combination high-level schedule and checklist list where tasks are identified and grouped on the schedule but no time is given for the subtask. Time is instead given to a group identifier within the high-level schedule.

Figure 20 is an excerpt from an ERP implementation schedule and defines a portion of the analysis phase to determine client requirements. Most of the tasks listed in this excerpt are simply checklist items. Rather than create a schedule using these items, they are simply listed. Subtask/ task list groupings have been created prior to schedule creation and assigned at tasks 8, 14, and 22. Task/checklist durations have been assigned to the relevant subtask and set at zero, while subtask groupings have been assigned 5 days each. Each 5-day term is representative of the amount of estimated time needed to accomplish all task list items listed for each grouping.

Notice that each checklist item has the same start date and that each start date for all checklist items aligns with the start date of the task list grouping. To accomplish this each checklist list item has been linked within the schedule to the task list subgroups. Checkmarks indicate those task list items that have been accomplished.

Task number 23 is the project buffer and was added during the critical chain scheduling process within ProChain Pipeline. In this way the project can be managed at a high level so that the project manager does not lose focus, while the individual subtasks are maintained as checklists.

C. Buffer Reports

As seen earlier, critical chain project management provides a number of methods for keeping projects on time. Reports are prepared that show the health, in percentage terms, of those buffers that may be affected by the lateness of a specific task or group of tasks, the objective being to focus activity on those tasks causing the delay. Additionally, reports are prepared to notify those resources that are approaching a task along the critical path so that they can be prepared to begin the task as soon as possible.

Included in the buffer report should be the buffer affected, the name/ number of the offending task or tasks, and the percent of buffer taken. Figure 21 is an example of a buffer report. Notice that the project and feeding buffers are from the house-building schedule discussed earlier.

ProChain Non-Resource Buffers as of 5/14/02 10:40 PM
REMSVI

ID	Buffer Name	% Buffer Used	Buffer Left	Chain Left	Orig Buffer	Orig Chain	Check Tasks
10	FB|Paint Exterior-12|Paint Interior-36	0 %	4 days	7d	4 days	7d	4-6;Build frame
16	FB|Install Plumbing-10|Install drywall-11	0 %	1 day	2d	1 day	2d	4-6;Build frame
20	PB|Install Fixtures & Appliances-19	0 %	9 days	17.25d	9 days	17.25d	4-6;Build frame

Figure 21 Nonresource buffer report. (Prochain® and Prochain Pipeline® are registered trademarks of the Prochain Corporation. All rights reserved. Screen shots reprinted by permission of the Prochain Solutions Corporation.)

The amount of buffer used is currently zero for all three buffers shown, meaning that no problems relative to the critical chain have yet to cause sufficient delay to reduce the amount of buffer available.

In Figure 22 a different condition exists. The feeding buffer located at task number 16 has been completely consumed. The feeding buffer at task number 10 has been 75% consumed, while the project buffer is 11% consumed. There are 14.25 days left in the critical chain and 8 days left in the project buffer. The tasks to be checked for problems that may have caused the buffer consumption include 6, 7, and 14.

To determine the overall health of the project the project manager should compare the amount of project buffer remaining with the amount of critical chain yet to be performed. Had the project buffer shown more than 50% consumption with over half of the critical chain left it would mean that the project would be in trouble.

The resource buffer report (Fig 23) gives a list of all tasks the resource buffers feed, the buffer start and stop times, the length of the buffer, and a notation of the number of days left until a task on the critical chain is to be performed. Task ID number 3 is a resource buffer feeding task number 4, build frame (see also Fig. 18). The resource being buffered is the framer. The buffer starts at 8:00 AM on the May 16 and ends at 5:00 PM on the May 17. The resource buffer length is 2 days. The number of days until the critical chain task is given as 0 days.

Note: Resource buffers are no longer recommended or required by current TOC standards. Practitioners have found that simple schedule reports are adequate.

D. Challenging the Assumptions

As seen earlier the first two steps of the improvement process are to level the load and to identify the critical chain. Figure 24 is an isolated view of the critical chain from the ERP implementation schedule seen earlier. The next step is to review the critical chain for those tasks that can be removed or shortened. There may be assumptions made that may not be valid. Break the assumption and the critical chain can be shortened.

The assumption chosen applies to task 6, gather business data. The assumption made is that task 2, conduct analysis preparation, and task 6, gather business data, cannot be accomplished at the same time.

To shorten the schedule a new assumption is formed. The new assumption is that analysis preparation and gathering business data can be accomplished simultaneously.

Figure 25 was created as a result of the new assumption. Notice that the start date for the task for gathering business data is much earlier than the original schedule and that it is no longer part of the critical chain. The result is

ProChain Non-Resource Buffers as of 5/27/02 11:31 PM
REMSM

ID	Buffer Name	% Buffer Used	Buffer Left	Chain Left	Orig Buffer	Orig Chain	Check Tasks
16	FB|Install Plumbing-10|Install drywall-11	100%	0 days	2d	1 day	2d	14-10:Install Plumbing
10	FB|Paint Exterior-12|Paint Interior-36	75%	1 day	7d	4 days	7d	7-18:Install Windows
28	PB|Install Fixtures & Appliances-19	11%	8.01 days	14.26d	9 days	17.25d	6-7:Roof house

Figure 22 Nonresource buffer report.

ProChain Resource Buffers as of 5/15/02 1:59 PM
REMSVI

Task ID	Buffer Name	Buffer Start	Task Start	Buffer Length	Days Until CC Task
3	RB|Framer|Build frame-6	5/16/02 8:00 AM	5/17/02 5:00 PM	2 days	0d
6	RB|Roofer|Roof house-7	5/20/02 8:00 AM	5/21/02 5:00 PM	2 days	2d
12	RB|Electrician|Install electrical wiri	5/22/02 8:00 AM	5/23/02 5:00 PM	2 days	
15	RB|Drywall Man|Install drywall-11	5/24/02 8:00 AM	5/27/02 5:00 PM	2 days	
18	RB|Carpenter|Frame/Finish Door J	5/29/02 8:00 AM	5/30/02 5:00 PM	2 days	
20	RB|Painter|Paint Interior-36	6/3/02 8:00 AM	6/4/02 5:00 PM	2 days	
23	RB|Floor Man|Install Flooring-35	6/6/02 8:00 AM	6/7/02 5:00 PM	2 days	
25	RB|Carpenter|Hang Doors-17	6/10/02 8:00 AM	6/11/02 5:00 PM	2 days	

Figure 23   The resource buffer report.

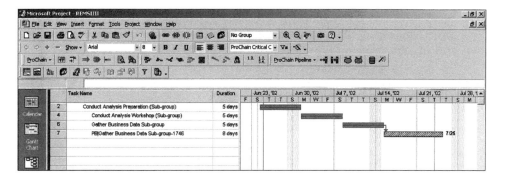

Figure 24 Identifying the critical chain to support the continuous improvement process.

that the completion date has moved from 7/25 to 7/18, a reduction of 5 business days.

Once the assumptions have been broken the project schedule is updated with the new project sequence and the process is repeated to find additional assumptions.

VII. CONCLUSION

Critical chain project management promotes better synchronization and decision support mechanisms, thereby providing better on-time, on-budget performance with more predictable due dates. Additionally, it provides an improvement process that significantly reduces lead times so that projects can

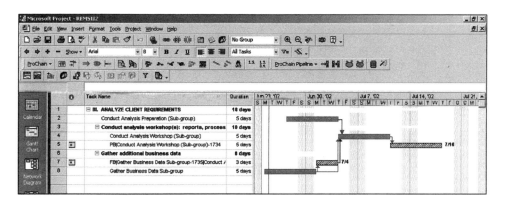

Figure 25 Breaking the assumptions—revised critical chain.

be finished earlier. Its focusing mechanisms and buffering systems insure that project managers focus their efforts and resources on those tasks that are critical to insuring on-time delivery.

VIII. STUDY QUESTIONS

1. What are the traditional reasons given for the failure of projects to reach on-time and on-budget objectives? What are the reasons provided by critical chain project management strategies? How are they different?
2. Define the critical chain concept and explain how it is different from the critical path method.
3. What is the major difference in buffering processes between the traditional critical path method and critical chain? What are the three types of buffers in the critical chain method? Describe how each is used and for what purpose.
4. What is the impact of multitasking, even for the most trivial of activities?
5. Describe in detail the critical chain project management continuous improvement process.

16
Scheduling the Multiproject Environment

Whenever more than one project is being scheduled and the projects have resources in common the probability is high that there will be resource contention between projects. Not only will each project be forced to contend for resource time, but resource time may also be limited for all projects. This brings up some interesting issues:

- How are physical constraints identified?
- How are they exploited?
- Does a subordination process still apply?
- How should multiple projects be scheduled?
- Will near-constraint resources interact to prevent or hamper scheduling?

I. CHAPTER OBJECTIVES

- To understand the multiproject environment
- To understand how to schedule multiple projects

II. MANAGING CONSTRAINTS IN MULTIPROJECT SCHEDULING

A. Identification

Unlike in manufacturing, project resources are usually knowledge workers. These knowledge workers may only have 50% of their time taken by project schedules. The rest is taken by nonscheduled tasks. If a delay should occur, the knowledge worker usually works overtime or transfer nonproject time to the project. So more than likely a physical constraint will not exist.

For scheduling purposes, critical chain uses the concept of pacing resources. A pacing resource is a resource that, while not necessarily being a constraint, has more demand placed on it than any other resource within the group of projects being scheduled. This concept would also apply to physical constraints. In critical chain, physical constraints are identified as pacing resources. Pacing resources can usually be identified in a number of different ways:

- The fraction of the resource's time that is modeled in the schedule
- The intuition of various people within the organization
- The perspective of each resource within the schedule

If resource A has 60% of its available time taken by a given schedule, while other resources are only loaded to 50% or less, there is a great probability that resource A should be treated as the pacing resource and, as such, be modeled to insure that other resources can be subordinated to it.

Note: The concept of modeling is simply the creation of a critical chain project schedule using the primary candidate as the pacing resource.

Additionally, project managers may have a sense of where delays are occurring, from a historical perspective, that prevent them from completing their tasks on time, or they may know who spends the most overtime trying to stay on schedule. Identifying a pacing resource may be as simple as asking the individual project managers or the resource managers themselves.

B. Exploitation

Interestingly, because of the tasks involved and the fact that resources within projects are usually scheduled to 50% or less, exploitation of resource time through the creation of a drum type schedule, as in manufacturing, has proven to be unproductive.

Instead projects are scheduled for completion as early as possible with each project being scheduled in succession. In other words, the first project date is scheduled based on due date, earliest due date first; then the next project in time is scheduled. Each project is scheduled to finish as early as possible.

A major difference between manufacturing and project scheduling is that project tasks rarely, if ever, have a setup. Attempting to maximize the utilization of the constraint by creating a schedule that saves setup would be a waste of time. Instead emphasis is placed on the continuous improvement process as discussed earlier.

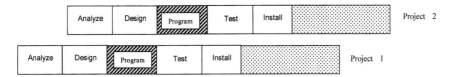

Figure 1 Generating the schedule for multiprojects.

C. Subordination

As with manufacturing, projects usually exist in the dependent variable environment and therefore are subject to issues regarding subordination. During the scheduling process, the pacing resource is subordinated to the completion date, and the remaining resources are then subordinated to the pacing resource. During the original scheduling process the relative position of each task is identified and attached to the succeeding task. When the multiproject schedule is created tasks are shifted relative to the pacing task start and stop times.

III. GENERATING THE MULTIPROJECT SCHEDULE

Multiproject schedule generation begins after each individual project has been scheduled and can be more effectively described as a consolidation process. Each task within all projects has its own relative due date and all buffers have been put in place. Consolidation begins after the first project has been placed in time. The next schedule is placed so that the pacing resource start time for the second project appears immediately after the stop time of the first project. In Fig. 1 programming has been identified as the pacing resource.

Two basic rules for the placement of successive projects are that none of the task times for previously placed projects is to be moved, and each project is to be completed as soon as possible.

In reality, resource contention problems are not the general rule and therefore, with exception of the pacing resource, contention problems are usually solved through buffers.

IV. CONCLUSION

Multiproject environments present some unique challenges not only for being able to maximize profitability for project management companies, but also in doing so while managing numerous projects. Critical chain multiproject capabilities minimize resource contention between projects while maximizing management visibility.

V. STUDY QUESTIONS

1. Define a pacing resource and describe its impact on multiproject scheduling.
2. How are pacing resources and resource constraints identified?
3. What key issues impact the exploitation of resources within the multiproject environment?
4. Define the subordination process for the multiproject environment.

Appendix
ERP Implementation at ABC Company

I. DESCRIPTION OF ABC COMPANY

With $1 billion in sales annually, four manufacturing facilities in various locations within the United States and distribution facilities located throughout Europe, Asia, the United States, and South America, the ABC Company manufactures medical implant devices and the instruments used during implantation. They have a wide assortment of products requiring precision machining of exotic materials and are a fast growing and quickly evolving business. Various companies are introducing new products into the market almost monthly.

The medical implant environment is highly regulated by the Food and Drug Administration (FDA) and requires that any production, environmental, procedural, or system changes be well tested and documented. Prior to introduction, any new products must meet FDA approval.

A. Processes

ABC Company creates ceramic moldings from wax fixtures; molten metals are then poured into the moldings and allowed to cool. Once cooled the moldings are machined to very tight tolerances. Special care is taken to ensure that any oils used in the production process are removed completely to avoid problems when implanted. After packaging, all products are bombarded with radiation to kill any bacteria that may be present.

B. Logistical Systems

The original ERP system was mainframe based and consisted of a mixture of highly modified commercial and homegrown applications, including material resource planning, shop floor control, bills of material, routings, purchase order management, general ledger, accounts receivable, and accounts payable. Forecasting was also a commercial system but was specialized and separate from the manufacturing system. The sales order management system was developed in-house.

Interface from forecasting to planning was via batch program. Finished goods were forecasted by distributors and consolidated as much as 6 months to 1 year in advance. As a general rule, inventory was sold from stock.

The mainframe system was to replaced by a server and a network of fat clients. Inventory consumption was to be managed by distributors and distribution centers via Internet connection. The mainframe ERP software was to be replaced by a completely integrated system. However, forecasting would still be a separate/specialized commercial system with a batch interface to planning.

C. Decision Processes

Traditional cost accounting permeated the entire company from product costing to sales order acceptance. Work centers were measured based on output per unit at each department. Bonus pay was given for personnel working at work centers or groups of work centers that exceeded standard. The incentive system for sales was based on sales price minus the cost of goods sold. Cost of goods sold was calculated as material plus labor plus overhead. Those products having the largest margin brought in the biggest commissions. Purchasing measured success based on whether or not the buyers met or beat standard.

II. ESTABLISHING THE PROJECT PHASES

It was obvious from the beginning that if the implementation was to be successful the executive committee needed to be educated so that they could properly set the ultimate objectives and the intermediate objectives for the implementation. When asked, all members of the committee had replied that they had already read *The Goal* and that it made great sense. However, a full understanding of how to convert the company to support the concepts and their impact would take an additional effort.

The initial education program for the executive committee consisted of

- A discussion of measurement systems and their impact
- The advantages of synchronization in the factory and in distribution
- An introduction to the profit methodology discussed in Chapter 14 of this book

Project Phases			
Phase	**Improvement Strategy**	**Location/Department**	**Target Date**
I	Factory Synchronization	Sales, Planning, Production, Production Control, Manufacturing Engineering	
II	Decision-Making/Support	Sales, Production, Inventory, Purchasing, Distribution	
III	Recession-Proofing	Sales and Marketing, Production	
IV	Strategic Planning	Corporate Office	
V	Add-On	All Applications/Departments	

Figure 1   Setting project phases for ABC Company.

After initial education steps, and with coaching, the executive committee decided that the company's options would be enhanced if the factory were synchronized. However, ABC had a very large distribution system. Fixing the synchronization process in the factory without addressing distribution would mean that a large portion of the benefits gained in the factory would be lost in the distribution system. It was imperative that the synchronization of distribution occur during or just after phase one. Because of the impact the committee decided to address both issues in phase one.

Phase two would concentrate on providing the necessary reports and applications to support the decision process, while phases three and four would focus on recession-proofing and strategic planning, respectively. Phase five was established to address any desirable functionality that was available in the software but had not been addressed in the first four phases. See Fig. 1.

III. PHASE ONE

A. Phase One Objectives

Because the market was growing at a relatively fast pace the executive committee felt that it would be possible to sell part of the excess capacity generated through synchronization by introducing new products that were already in the FDA approval process. While this revenue had already been accounted for in the forecasts it would offer the company an opportunity to generate more revenue with less risk by not increasing operating expense.

While finished goods were made to stock, distributor forecasts were notoriously wrong and would require multiple changes during the year. In fact, planning would often ignore the forecasts and build what they wanted to build. The shorter lead times due to synchronization would mean more accurate forecasts that, in turn, would mean less inventory inaccuracy in finished goods. The committee decided to pass the shorter lead times throughout the distribution system including distributors and hospitals. The impact would be that distributors would be better able to compete.

Additionally, because of the cyclical nature of the business, ABC had been forced to build ahead of demand during slow periods so that they could continue to provide product during times of peak demand. By increasing capacity the company could reduce the amount produced during slow periods and increase the amount produced during the peak periods. By so doing, there would be a major impact on finished goods as well as work-in-process inventory.

The committee felt that if production could cut its lead times in the factory by 25%, throughput would increase by 15%. A better synchronization of the factory also meant a reduction in inventory. Current estimates were set at between 30 and 40%.

The reduction in overtime associated with the increase in available capacity and a reduction in carrying cost associated with the lower inventory would result in operating expense dropping by 8%.

Synchronization of distribution would insure that the 15% increase in throughput generated in the factory would be realized as throughput entering the company from the distributors and also result in an additional reduction in lead time through the distribution center to the distributors. As lead times were reduced so would be the requirement to store buffer inventories at the distribution centers. ABC estimated that distribution inventories would drop by as much as 40%. See Fig. 2.

B. Phase One Intermediate Objectives

The next step was to determine what was necessary to synchronized the factory and the distribution systems (the intermediate objectives) and document the profit improvement plan. To facilitate this a project manager was identified and, as a result, the project team began to take shape. A war room was also set up in one of the company's conference rooms to house a number of computers, the project team members, and education/training sessions.

Phase 1 Objectives	
1	Increase throughput by 15%
2	Reduce factory inventory by 30%
3	Reduce distribution inventory by 40%
4	Reduce operating expenses by 8%

Figure 2 Establishing the phase one objectives.

Traditional ERP systems, as documented earlier, are incapable of supporting factory synchronization. So the implementation must include some plan to develop this kind of capability. To be effective, the implementer must have the ability to (1) place orders on a timeline for the constraint (the schedule), (2) provide some type of buffering mechanism for those things that will go wrong, (3) correct the system when things to do wrong, and (4) provide a release mechanism for the gating operations. Additionally, (5) assurance must exist that employee behavior and management measurement systems would not sabotage the implementation.

Traditional distribution systems and policies have the tendency to optimize inventory distribution throughout the supply chain resulting in an increased amount of inventory, a lengthening of lead times to the customer, and an extension of the forecast horizon. The longer the forecast, the more inaccurate it becomes. To be effective the implementer must (6) eliminate the policies that support local optimization of the supply chain.

These six issues proved to be the biggest obstacles for the project team and were selected as the intermediate objectives for phase one.

The intermediate objectives needed to be accomplished very quickly so that the time to payback could be minimized. The choice facing the team was to either buy additional functionality of the type described in this book or attempt to modify the ERP system selected earlier. Rather than spend additional money for software that had not been in the original budget the project manager decided to investigate the possibility of modification.

Fortunately, the project manager was able to find experienced help in this area and accomplished the initial design very quickly. Much of the functionality required was already in the standard ERP software selected. The modifications required proved to be minor in nature and consisted of the creation of a few batch programs, reports, inquiry/application screens, and some files. As luck would have it, some of the standard files had fields that were not currently being used, but would provide a means for finite scheduling. While the functionality and flexibility would not be as great as that found in an off-the-shelf TOC-compatible system, the objective for the project manager was to accomplish the phase one goal.

One of the toughest issues faced by the project team was changing the bonus program for workers. The bonus programs actually created dysfunction within production and desynchronized the factory. Eliminating the way the bonus system worked was a must but proved to be a very touchy subject for employees as well as management. The employees did not want to lose their bonuses and management was leery about what might happen when the employees were told about the new policy. A decision was made to increase the pay of production workers to offset for the pay loss due to the elimination of the bonus system. Since the same amount of payroll would leave the company regardless of how the employees were paid there would be no impact to operating expense.

C. Prerequisites and Application Relationships

After the system requirements had been established, the team was ready to define the prerequisites and application relationships. The stipulation given by the project manager to the project team was that the manager must approve any feature over and above the minimum to meet the intermediate objectives for phase one. This invariably led to some interesting conversations about what was required. For instance, were requisitions required before purchase orders could be released; were advanced pricing or blanket orders necessary for phase one; were receipt routings also necessary; would the implementation include invoicing?

One of the key issues was that the team consisted of users. The users were required to create what was needed. This eventually resulted in the users taking ownership of the end product—so much so that they were ready to defend it exuberantly during user training.

Members of the team included people from sales, production, planning, purchasing, inventory, and engineering. Also included were two developers, one for the required modifications and the other for data conversion, and two consultants. Each consultant acted as a coach to the project team for an area of specialty, one for manufacturing and one for distribution.

Education and training for the project team were customized to focus on the phase one requirements. Additional software functionality was eliminated from the course curriculum, and the team focused only on the required data and applications necessary for factory and supply chain synchronization. Education focused on helping the team understand factory and supply chain synchronization and preceded training. Education subjects included cause-and-effect relationships in the dependent variable environment, drum-buffer-rope methodology, and understanding the decision process. Training included excerpts from the vendors training guide. Subjects included sales order management, purchase order management, shop floor control, material planning, inventory management, and product data management.

D. Phase One Profit Improvement Plan

Once all the application requirements were defined the project manager was ready to document the profit improvement plan for phase one. The total investment over the life of the project including software, hardware, and consulting was estimated to be $8 million. However, payback was due to begin after the first phase rather than at the end of the project, and while there was no difference in the layout of expenditures for the software and hardware from a traditional implementation, the estimated $2.5 million in consulting fees would be prorated over the life of the project. Based on the project objectives break-even would be 3 months after phase one go live. The entire project would take 12–18 months. This meant that the project would be paid for before it ended.

E. Phase One Implementation

The schedule for the implementation included a rollout at ABC's Texas facility and was due to begin within 6 months of the beginning of phase one. Once the system had been configured and the conference room pilots (CRPs) were complete, user training and go live would follow 1 month apart until all facilities had gone live.

   According to the schedule, the system needed to be completely configured and all modifications made by the end of the end of the fourth month so that a stand-alone conference room pilot could be performed. During the CRP each module would be tested for its functionality and usefulness in attaining the phase one goal. All issues were documented and later addressed to solve any problems that the team encountered.

   An integrated conference room pilot was schedule for the end of month 5 to test any changes from the first CRP and to determine how well each module interfaced with the others. Because of the commitment of the project team both schedule dates for the conference room pilots were made on time.

So that users would not forget what they had learned, the user training schedule was set so that users could be trained and then immediately begin to work with what was called a "sandbox environment." This sandbox environment was a complete duplication of the CRP environment so that users could practice what was learned in training. Workers were encouraged to stay on site an extra hour each day to work with the system.

While some used the sandbox environment others were put to work preparing the production environment for go live. Most of the data were to be converted from the legacy system, while other data were to be loaded manually. Key issues included purchase orders, sales orders, bills of material, routings, and work orders. Additionally, after conversion the system was to be tested to insure that all data were converted successfully.

Preparation for go live was also made on the shop floor. Areas where inventory would accumulate due to mismatched parts or overutilization of a particular machine were moved to positions in front of the constraint. A label was placed on the floor designating these as buffer zones.

So that orders would be processed in a first in, first out (FIFO) fashion a list was maintained noting the date and time that an item entered the buffer area.

New inventory levels at distributors and distribution centers were reset based on considerably reduced replenishment lead times.

F. Phase One Go Live

The go live process included a 2-week handholding and problem-solving exercise with the ERP consultants on site to help with any configuration changes that might be necessary. Each transaction was tested on live data, and problems encountered were fixed prior to being released for day-to-day operations.

While sales, purchasing, and inventory processes were straightforward traditional implementations the synchronization process was not. Part of the pre–go live process was to drain excess inventory from production. Workers were not allowed to build ahead, process orders out of sequence, or release orders early.

During the go live process finite schedules were released for the constraints, and checks were made by the buffer managers to insure that materials needed for processing were at the constraints. If not, the material was expedited. In this case, because of the previous attempts to limit the amount of overutilization of equipment, expediting was held to a minimum.

Buffers were initially oversized to insure that the constraints would have enough work. Once a certain comfort level was reached with the synchronization process the buffers were reduced in size and monitored by the buffer manager.

G. Phase One Result

The results of the phase one implementation were quite good. Finishing on time, the estimated 30% drop in inventory at the factories actually hit closer to 50%. Throughput increased 16%, while operating expense decreased by 12%. Lead time to the distributors and hospitals was cut in half, while distribution inventories were reduced by 60%. The phase one goals were actually exceeded.

IV. PHASE TWO

A. Phase Two Objectives

By far the most interesting phase of the entire project was phase two. While phase one was primarily a production responsibility with data being created by other departments, phase two would result in change occurring throughout the company. This would take a much greater effort to accomplish. Among the reasons for this are

1. While the goal of the phase must be to increase the profitability of the company, changing the decision process can only increase the probability of success. Unless the constraint (or constraints) to making more money is exploited or elevated additional throughput will not be realized.
2. If the constraint is not identified and the impact of its exploitation or elevation is unknown, then the profit improvement plan cannot be written, and a justification for phase two would not exist.
3. If a surgical approach is used that isolates the constraint (or constraints) to making more money and then exploits or elevates it, the project will probably take forever (or maybe just shy of that date) to finally conclude. The longer the consultants are on site, the more operating expense will be incurred by the company.
4. The alteration in how people think about their company and the measurements they use will be a major change in what they previously viewed as reality. To be effective, the implementation plan must include a method of handling this kind of change.

In addressing the first and second issues, there are two ways to increase the probability that an increase in profitability will occur when a decision process is changed. One is by using the TOC thinking process to identify the constraint (or constraints) and then exploit or elevate it. The other is by identifying those decision processes that will have a high degree of probability of creating higher profits. Take product mix as an example, if a company is using traditional cost accounting, there is an excellent possibility that its current product mix does not maximize throughput and that changing the decision process to include throughput per unit of the constraint instead of sales price minus cost of goods sold would have a major impact.

While some decision processes may have a high probability of an immediate impact to profitability, others do not. If the decision process for justification was changed from cost accounting to throughput accounting and there are no purchase requests or requisitions to buy equipment, then there would be a low probability (or zero) that an implementation of throughput justification would increase profits near term. However, not implementing them may result in profit losses in the future. If someone requisitions new machine in the future and through-put accounting has not been implemented, then there is a good chance that operating expense will increase without a corresponding increase to throughput.

ABC Company chose to attack all decision processes at once. It seems that, to the executive committee, there was no great benefit in implementing a bad decision process and then changing it later. The only caveat was that those decision processes having an immediate impact to throughput and those applications that could be interpreted as necessary conditions should be identified and given the highest priority. The executive committee also wanted to see a profit improvement plan prior to implementation so that a go/no go decision could be made.

This meant that every decision process that the company used on a day-to-day basis needed to be reviewed and a valid replacement found. To prepare for the implementation, the decision processes were divided into two different categories: those decision processes having a high probability of increasing profit immediately and those having a low probability. Those decision processes having a high probability, in addition to finding replacements, were reviewed to quantify the impact of the change to throughput, inventory, and operating expense so that the project objectives could be set and the profit plan written.

In addition to identifying the decision processes and their impacts, the company needed to determine what data/applications would be in effect necessary conditions such as advanced pricing, adjustments, or rebates.

A major point of the decision about fulfilling necessary conditions was that if a question arose about the inclusion of a specific application and the process could be handled manually or with the legacy system, then it was delayed to phase five.

The project team identified a number of decision processes that, if implemented successfully, would result in a relatively quick improvement to profitability and could be linked directly to the actions taken to attain the project objectives. They included but were not limited to

- Product mix
- Sales order acceptance
- Finding and selling excess capacity
- Renewed focus on quality, engineering, and maintenance in the factory

In making a determination of necessary conditions, the team identified a number of applications that needed implementing in phase two, including

- Electronic commerce
- Product cost accounting

The huge amount of sales orders being processed daily made it necessary for customers to have the ability to enter sales orders from the field. An implementation of electronic commerce would provide the ability for distributors to enter sales orders through the Internet to consume the forecast. Additionally, a requirement of the Generally Accepted Accounting Principles (GAAP) made it imperative that the company value inventories in a specific way. This included a traditional cost rollup of material, labor, and overhead. With this information the team began to build the project's profit objectives.

By changing the company's product mix at each production facility the project team calculated an increase in throughput of 5%. Changing the way sales orders were accepted from the profit margin (sales price minus cost of goods sold) to throughput per unit constraint would increase throughput by 5%. Finding and selling additional capacity was estimated to add another 4%. A refocus of the quality, maintenance, and engineering efforts would increase throughput by 2%. Inventory and operating expense were expected to remain level.

The phase two objectives are reflected in Fig. 3.

Those decision processes that had a low probability of creating an immediate improvement in profit included but were not limited to

- Purchasing—what to pay and make versus buy decisions
- Finance—cost justification
- Production—Scrap versus rework

While these processes were not included in the development of the phase two profit objectives it was important to include them so that they would not become constraints in the future.

Phase 2 Objectives	
1	Increase throughput by 16%
2	Inventory to remain level
3	Operating expense to remain level

Figure 3   Establishing the phase two objectives.

B.   Phase Two Intermediate Objectives

Again, the intermediate objectives would identify those actions necessary to attain the phase objectives. They included the creation of reports and inquiries to aid in the decision process. Added were the specific classes designed to insure that everyone understood the necessity and impact of the change. The classes also included excerpts from the vendors training classes that pertained to any new functionality needing to be implemented to support the phase two actions and necessary conditions.

Additional intermediate objectives, obstacles, and actions were set for the rest of the decisions that the ABC project team had identified. Like the first set, classes were created to support the changes required. See Fig. 4.

C.   Prerequisites and Application Relationships

Once the prerequisites (the data used in the decision process) to the actions were identified a relationship was built that would identify the applications that must be included in the implementation to enable ABC Company to make the required decisions. Most of the applications had been implemented in phase one. However, two more were required:

- Advanced pricing and adjustments.
- Product cost (raw material)

In order to calculate the amount of throughput to be gained by a certain decision the decision-maker needed to apply the cost of raw material to the sales price. The product cost application would provide the means to roll the

Phase 2 Intermediate Objectives		
Int. Objective	**Obstacle**	**Action**
Product MIx	No product mix Information	Create reports and Inquiry
	Current process cost based	Break cost assumptions
Sales order acceptance	No S/O acceptance inf.	Create reports and inquiry
	Current process cost based	Break cost assumptions
Find/Sell excess capacity	Identifying capacity/throughput chains	Create reports and Inquiry
	Current process cost based	Break cost assumptions
Quality/Eng./Maint. focus	No focus information	Create reports and Inquiry
	Current process cost based	Break cost assumptions

Figure 4   Establishing the phase two intermediate objectives.

cost of raw material into the product. Of course, product cost was also included in the necessary conditions defined earlier. Because of the way the vendor's software worked and because of the file structure, it would be easy to isolate the raw material cost from each of the cost types used for the total cost required by GAAP. Advanced pricing would provide the sales price.

After the applications were identified the project team was able to create a schedule and estimate the amount of additional operating expense needed for implementation. At that point a break-even could be established.

D.   Phase Two Profit Improvement Plan

Break-even included the comparison between consultant/developer fees and the profit objectives for phase two. The cost of software and hardware had not been included in the comparison because each was to be paid for by phase one. The development/configuration portions of phase two were projected to last for 4 months at a cost of $450,000 and begin generating benefits within the first month after phase two go live. The break-even point was set at 2 months after phase two go live.

Note: Because of the increased throughput generated by phase one, the entire project was actually paid for in the fourth month after phase one go live.

E. Phase Two Implementation

Configuring the additional applications used to support the decision processes and developing the reports and inquiries were all accomplished from
ABC's Texas facility. Go live was to be accomplished in the same manner as
phase one except that timing was reduced from 30- to 15-day intervals for the
rollout at each branch plant.

The loading of advanced pricing data began during month 2 of the
implementation after a pricing strategy was determined and data made
available. The new pricing strategy would take advantage of the new decision
processes and capabilities.

All development and configuration needed to be complete by the end of
month 3 so that the CRP could be accomplished with enough time to make
any modifications required and to conduct all necessary training.

A combined (stand-alone and integrated) CRP was used to test individual applications and also the interfaces to other modules. Each report/
inquiry was used to make a decision and the impact of the decision tested to
insure that, if done correctly, a positive impact would be made to profitability.
Changes were made to raw material costs, pricing, routings (constraint time),
and bills of material to insure that the changes were reflected in the reports
correctly. Advanced pricing was tested to insure that the correct pricing
schedules were used and adjustments made when sales orders were added or
changed. Sales order entry via electronic commerce was tested to add, change,
and delete orders.

A sandbox environment was again made available for users right after
the training classes.

F. Phase Two Go Live

Like the phase one go live, an initial testing procedure was used to test each
transaction, report, and inquiry in the production environment. The test was
to insure that all programs, versions, and menus were promoted correctly and
that they worked. Rollout of electronic commerce for distributors and
hospitals was an extended process due to the massive number of locations
(approximately 1400 throughout the world).

G. Phase Two Result

Phase two resulted in an increase in throughput of 17%. Inventory dropped
during the phase two go live as a continued improvement from phase one and
due to the refocus program. Interestingly, the refocus program made a larger

impact on lead time, inventory reduction, and operating expense than it did on throughput. Finding and selling excess capacity and changing the product mix were the biggest surprises. Rather than increasing throughput by the combined estimate of 9%, throughput increased by 12% for both efforts. Operating expenses actually dropped by 5%.

V. PHASE THREE

The method used in the recession-proofing process was to maximize the number of throughput chains that ABC could support using the same resources and to insure that all markets associated with the chains would not be in recession at the same time. To do this the company needed to be able to identify existing chains, identify where excess capacity existed, and define new products or modify existing products to create new chains using the excess capacity. The ability to identify the throughput chains and excess capacity had already been created during phases one and two. However, trying to determine the impact of adding new throughput chains for products or chains that do not yet exist is best done through simulation. This was an important issue that would more than likely have been solved with off-the-shelf TOC compatible software. It took much longer than originally thought to accomplish recession-proofing.

VI. PHASE FOUR

ABC Company had undergone tremendous change in a relatively short period of time and, as a result of phase three, was running critically short of capacity. Their current facilities were bursting at the seams. So a plan was devised to build a new facility.

Note: The market segmentation and recession-proofing portions of strategic planning were included in phase three. Phase four involved company growth and long-term planning. Once again the company was hampered by the lack of a simulation capability.

The plan adopted by the executive committee included transferring enough throughput chains from each of the existing facilities to the new facility so that it would be viable. Also, building a new facility meant that a new implementation of the information system would be required. Once the selected products were transferred, all facilities began the recession-proofing process again.

VII. PHASE FIVE

Phase five of the implementation ran concurrently with phases three and four and included the addition of those applications not originally planned for phases one and two. The rest of the implementation included but was not limited to

- Cycle counting
- Purchasing requisitions
- Requisition approval
- Receipts routing
- Sales order hold/release
- Purchase order quotes

Since no increase in profit was expected from phase five no profit objective was set. Most of the additional transactions only needed minor setup and were ready for go live 2–3 weeks after the end of phase two. An abbreviated conference room pilot and training session were conducted at the end of week 3, and go live was conducted the following week. Rollout was accomplished in 1-week intervals.

VIII. SUMMARY

From start to finish the ERP system implementation took 12 months to implement through the end of phase two. Profits were up drastically. Within 14 months of the start of phase one ABC increased throughput by over $20 million per month. The entire project was paid for by the first week in month 10.

Glossary of TOC Terms

A-plant The A-plant is characterized by a large number of converging operations that start with a wide variety of raw material items being assembled in succeeding levels to create a smaller number of end items.

Action An activity defined in a transition tree which is used to obtain an intermediate objective.

Activation The employment of nonconstraint resource for the sake of keeping busy. Activation is unrelated to whether it is useful in supporting system throughput (APICS).

Additional cause In the TOC thinking process, an additional cause is used when the presumed cause by itself is insufficient to explain the existence of an effect.

Aggregation of demand The result of the use of the formula capacity divided by demand in which all demand is aggregated into time buckets. The assumption is made that all capacity is available to all demand within the time bucket.

Artificial demand A feature of the dynamic buffering process, artificial demand originates from internal processing logic. It is demand placed on a resource to protect available capacity when a certain number of consecutive periods on a resource have been schedule at 100% by external demand.

Ascending part/operations list A feature of the system which lists individual part/operations in ascending order from raw material to the final product.

Assembly buffer Construct used to insure that those assembly operations directly fed by the constraint do not wait for material to arrive from nonconstrained legs within the net. The assembly buffer also determines the release schedule for raw material into those operations which feed the assembly buffer.

Assembly schedule   A schedule used to sequence the arrival of parts to an assembly operation from nonconstraint operations. Assembly schedules are only used when at least one leg of the assembly operation is being fed by the constraint(s).

Backward rods   Rods which require an order to be schedule across a constraining resource at a given amount of time after a previous order on the same or a different resource.

Batch rods   A time mechanism used to separate/protect orders (batches) on the same resource schedule.

Behavioral constraint   A behavior that is in conflict with reality and blocks the exploitation of a physical constraint.

Bottleneck   A resource is a bottleneck when the demand placed on a resource is equal to or more than capacity.

Budget   A performance measurement, the budget is used to relate the forecasted expenditure requirements at the department level. This information is used to predict and control operating expense and should not be used to focus improvement.

Buffer   A time mechanism used to protect those portions of the factory (buffer origins) that are vulnerable to problems associated with statistical fluctuations.

Buffer consumption   A buffer is consumed when the amount of time allocated for those things that can and will go wrong, Murphy, is reduced in relation to the amount of time required to produce.

Buffer management   A technique used to manage the amount of protection necessary and the process of controlling the buffer origins within the plant.

Buffer management report   The buffer management report consolidates the input of the buffer management worksheet and determines the relative impact by injecting the throughput dollar day and inventory dollar day equation.

Buffer management worksheet   Organized to facilitate quantitative analysis and to insure uniformity in data collection, the buffer management worksheet is used by the buffer manager to collect and assemble data relevant to those orders which require being expedited.

Buffer manager   The person designated to manage the buffer and insure that the buffer origin is protected by expediting material causing holes in zone I of the buffer.

Buffer origin That portion of the system that needs protection and is the object of the buffer. Buffer origins include, but are not limited to, the constraint, the secondary/tertiary constraint(s), shipping, and those assembly operations having one leg being fed by the constraint.

Buffer system The total system used to protect the constraint from the impact of statistical fluctuations.

Buffer zones Refers to the three zones within the buffer management process. (*see also* **Tracking zone** and **Expedite zone**)

Capacity Usually expressed in hours:minutes or quantities for a given period of time, capacity is the capability of a resource to react to demand.

Capacity-constrained resource (CCR) A resource which has, in aggregate, less capacity than the market demands.

Categories of legitimate reservations Used in the TOC thinking process, the categories of legitimate reservation question the existence of a specific entity, its cause, or the relationship between them.

Cause insufficiency Used in the TOC thinking process to define a situation in which the suspected cause is insufficient to explain the result.

CCR (*see* **Capacity-constrained resource**)

Combination plants Manufacturing plants which have more than one of the VAT characteristics. (*see* **V-plant**, **A-plant**, and **T-plant**)

Combination rods Rods which require an order to be schedule across a constraining resource a given amount of time before and after a previous or succeeding order on the same or a different resource.

Combination schedule The technique of combining high level task lists or schedules with checklists.

Constraint Anything which limits the system from attaining its goal. Constraints are categorized in the following manner:

- Behavioral
- Managerial
- Capacity
- Market
- Logistical

Constraint buffer Used to protect constraint(s) and to determine the release of raw material to those operations which feed the constraint.

Constraint management   The practice of managing resources and organizations in accordance with theory of constraints principles (APICS).

Constraint schedule   The schedule created for capacity-constrained resource in order to exploit its productive capability.

Continuous improvement (project related)   The four-step process designed to continuously reduce the length of the critical chain. The steps include

1. Level the load.
2. Identify the critical chain.
3. Challenge the assumptions of the critical chain.
4. Change the project to reflect the changes in assumptions.

Continuous profit improvement   The five step process designed to continuously improve the profits of the company. The steps include

1. Identify the constraint.
2. Exploit the constraint.
3. Subordinate the nonconstraints.
4. Elevate the constraint.
5. Repeat the process.

Converging operations   Operations which receive parts from more than one part/operation.

Cost mentality   The tendency to optimize local measurements at the expense of global measurements.

Critical chain   The name given to the combination of the critical path and the solution to resource contention within a project schedule.

Critical chain project management   The name given to a group of TOC-based project management strategies and rules.

Currently reality tree (CRT)   Used to find the core cause or causes from undesirable effects (UDEs).

Dependent setup   Used to maximize the utilization of the constrained resource by combining setups on orders for two different part/operations having the same setup requirement.

Dependent variable environment   The environment where resources are dependent on each other in their capability to produce and are subject to variation in that capability.

Descending part/operations list A feature of the system which lists individual part/operations in descending order.

Desirable effect Used in the TOC thinking process in the creation of the future reality tree, refers to those effects which are beneficial and are created as a result of an injection.

Diverging operations Operations which deliver part to more than one part/operation.

Drum The schedule for the primary constraint, establishing the rate at which the system generates throughput.

Drum-buffer-rope A scheduling technique developed using the theory of constraints. The drum is the schedule for the primary constraint and establishes the rate at which the system generates throughput. The buffer is a time mechanism used to protect those places within the schedule which are particularly vulnerable to disruptions. The rope is the mechanism used to synchronize the factory to the rate of the constraint and determines the release date for material at gating operations.

Dynamic buffering A method used to improve the buffering process so that overall buffer sizes can be shrunk. The buffer is allowed to grow when increases in resource demand require additional lead time. Dynamic buffers shrink as demand declines.

Dynamic data Data which are allowed to change as the environment and reality change.

Effect-cause-effect A method formerly used in the TOC thinking process that looked for core cause by continually establishing cause-and-effect relationships and supporting each relationship with an additional effect.

Elevating the constraint The act of increasing the ability of the constraint in the creation of throughput by means other than exploitation.

Entity Used in the TOC thinking process, the entity is used to identify either a cause or an effect.

Entity existence Used in the TOC thinking process, one of the categories of legitimate reservation which questions the existence of an entity.

Evaporating clouds (EC) Used to model the assumptions made which block the creation of a breakthrough solution.

Excess capacity   Capacity that is not used to either produce or protect the creation of throughout.

Expedite zone   Zone I of the buffer. The expedite zone is used to indicate when the absence of an order is threatening the exploitation of the constraint and to notify the buffer manager when to expedite parts.

Exploitation   The process of increasing the efficiency of the constrained resource without obtaining more of the resource from outside the company. Exploitation may include, but is not limited to, the implementation of statistical process control, total productive maintenance, setup reduction or savings, creating a schedule, and the application of overtime on a constraint resource.

Feeding buffer   A buffer used whenever a noncritical path task within a project feeds the critical path.

First day peak load   A load placed on a resource during the subordination process which exceeds available capacity on the first day of the planning horizon.

Forward rods   Rods which require an order to be scheduled across a constraining resource a given amount of time prior to a succeeding order on the same or a different resource.

Future reality tree (FRT)   Used to model the changes created after defining breakthrough changes from the evaporating cloud.

Global measurements   Those measurements used to measure effectiveness from outside the company (e.g., return on investment).

Glue   A technique used during setup saving to attach orders together for the purpose of scheduling the constraint. Once orders have been attached they remain attached until the order is complete.

"Haystack-compatible" system   A term referring to an information system based on the theory constraints and inspired by *The Haystack Syndrome* by Eli Goldratt.

High level schedule   A project schedule that eliminates detailed tasks and is used to determine the health of a project.

Holes in the buffer   A phrase that refers to the absence of orders within zones I or II of the protective mechanism called the buffer.

Inertia   The tendency to continue looking at a specific problem or activity the same way even though the situation has changed.

Injection Used in the TOC thinking process, the injection is the change used to eliminate undesirable effects.

Intermediate objective In the TOC thinking process, an objective attained to overcome an obstacle defined in the prerequisite.

Inventory One of the key measurements used to manage a TOC company. It is defined by Eli Goldratt as "all the money the system invests in purchasing things the system intends to sell."

Inventory dollar days A measurement used to minimized those activities which occur before they are scheduled. It is represented by the formula inventory quantity multiplied by dollar value multiplied by the number of days early.

Leg A term used when referring to the product flow diagram or net. A leg is a string of part/operations which feeds one part to an assembly operation. By definition an assembly operation will have two or more legs.

Leveling the load A phrase used to describe the initial attempt at scheduling the constraint. The objective is to place each order on the timeline so that there are no conflicts between orders and so that it does not exceed daily capacity.

Local measurements Performance measures used at the department level to guide activity.

Logical implementation One of the three phases of implementation of the TOC-based system, the logical implementation involves changes in the way in which the company is managed.

Logistical constraint Occurs when the planning and control system prevents the company from attaining its goal.

Managerial constraint A management policy that prevents or restricts the company from attaining its goal.

Market constraint Occurs when market demand is less than the capability of the company to produce.

Market segmentation A strategy for recession-proofing and maximizing profit by creating different markets to adsorb capacity from the same resource base.

Master production schedule Used as input in the creation of the net, the master production schedule is the combination of the forecast and sales order requirements. It provides the initial schedule by which the identification process begins to find the primary constraint.

Moving in time Unlike the low level code used to control the order of processing in material requirements planning, the TOC system uses time. *Moving in time* is a phrase used to describe the procedure used to process orders on the net. As each order is processed it is placed in time beginning with the end of the planning horizon. Once all activities have been planned for a specific date the system moves in time to pick up those requirements of the preceding period. Planning begins at the end of the planning horizon and moves in time toward time zero.

MPS (*see* **Master production schedule**)

Multiproject scheduling The consolidation of more than one project into a single schedule.

Multitasking The act of performing more than one task at the same time.

Near constraint A resource that is loaded to near capacity levels. While it may not limit the amount of throughput that the system can generate, it can have a negative impact on the subordination process and will result in an increase in inventory and operating expense.

Necessary condition Boundaries placed on a company's departments or individuals, originating either internally or externally, which serve to regulate activity.

Negative branch Used in the TOC thinking process, the negative branch is the reason why a certain activity cannot or should not be done. Negative branches are normally trimmed so that solutions become more robust.

The net The combination of all product flow diagrams for all products on the master schedule.

Nonresource buffer A buffer used in conjunction with a task in a project.

99-to-1 rule The rule stating that in a dependent variable environment, only 1% of the activity is required to produce 99% of the impact.

Nonconstraint A resource that has more than enough capacity to meet the market demand.

Nonconstraint schedule A schedule produced for nonconstraint resources presenting those activities which must not be done prior to a specific time.

Obstacle Used in the TOC thinking process, the obstacle is defined as something that will negate or reduce the effect of the injection and its impact on the undesirable effect.

Off-loading The process of removing specific orders from a constraint resource and placing them on a nonconstraint resource.

Operating expense One of the key measurements used to manage a TOC company. It is defined by Eli Goldratt as "all the money the system spends turning inventory into throughput."

Optimization The act of maximizing local measurements.

Overtime schedule The schedule for overtime requirements resulting from the exploitation process. The schedule includes the resource on which the overtime is to be spent, the application date, and the amount of time required.

Pacing resource The resource chosen as the focal point for creating the multiproject schedule.

Part/operation A combination of the part number from a bill of material and the operation within a routing that is used to produce a specific part. It represents the smallest element within the product flow diagram.

Peak load A period in time when the demand placed on a specific resource exceeds capacity.

Permanent education program An education program designed to support the implementation and execution of the theory of constraints.

Physical implementation That part of the implementation involving the interface of the user with the system, which includes plant layout, schedule generation, schedule execution, and buffer management.

Prerequisite tree (PT) Used in the TOC thinking process, a flowchart that uncovers and solves intermediate obstacles to achieving goals.

Process batch size The number of parts processed on a resource without intervening setup.

Process of continuous profit improvement The five-step process used in the theory of constraints consisting of identification, exploitation, subordination, elevation, and repeating the process.

Product flow diagram A diagram made from the combination of a bill of material and routing for a single product. It presents individual part/operations and their relative positions within the production process.

Productive capacity   That capacity required to produce a given set of products, not including that capacity used to protect the schedule from natural fluctuations in capability.

Profit improvement plan   The plan used to document the profit/breakeven goals of an ERP implementation.

Profit methodology   The strategy and rules applied to an implementation of an ERP system that focuses on achieving profit.

Progress reporting   In traditional terms, the act of reporting the percent complete of a task, in TOC terms this means reporting the amount of time remaining to complete a specific task.

Project buffer   The primary buffer extending from the due date of the project to the last task in the project.

Protective capacity   That capacity used to protect the capability of a resource or group of resources in meeting the schedule.

Pushing the load   The process of shifting resource load into later periods.

Pushing the order   The act of pushing a specific sales order on the schedule into a later time period.

Raw material schedule   The schedule produced by the system that defines the quantity and date of raw material requirements.

Recession-proofing   A methodology designed to ensure that the company will continue to grow regardless of the economic environment in which it exists. Included in the recession-proofing process is the ability to

- Continuously implement the five-step process of improvement
- Minimize the impact of a recession by protecting resources through proper market segmentation
- Make valid decisions

Red lane   The string of resources that lead from the primary constraint to the sales order. The red lane is the most vulnerable portion of the factory.

Release schedule   The schedule for the release of raw material into the gating operations.

Reservation   (*see* **Categories of legitimate reservation**)

Resource activation rules   The rules used by a resource to determine what to do and when to do it.

Resource buffer   A time buffer used to warn resources when they are scheduled to work on a critical chain task.

Resource contention   Whenever more than one task is scheduled to be performed by a resource during the same time period.

Road map   (*see* **TOC thinking process**)

Rods   (*see* **Time rods** and **Batch rods**)

Rod violation   Occasion when a specific order having forward, backward, or combination rods is compromised by a schedule in which another order occupies space too close in time.

Rope   The rope is the mechanism used to synchronize the factory to the rate of the constraint and determine the release date for material at gating operations.

Schedule conflict   A situation in which part/operations or batches scheduled on a given resource are in conflict with other part/operations or batches due to a lack of adequate protection.

Secondary constraint   A resource whose capacity is limited to the extent that it threatens the subordination of the primary constraint.

Setup savings   The act of combining orders to reduce the number of setups or time spent for setups and thereby increase the amount of protective or productive capacity.

Shipping buffer   Used to protect the shipping of finished goods, determine the initial schedule for the constraint, and establish the release schedule for raw material which does not go through the constraint or assembly buffer.

Shop floor control   That activity associated with controlling the production process, including (1) controlling the priority of work (the schedule for the constraints, the release of orders to the gating operations, the schedule for the assembly buffer, the schedule for the shipping buffer, and the buffer management system); (2) collecting and feeding order movement, location, and quantity information back to the shop floor control system; and (3) measuring tracking (collecting and reporting data for maintaining throughput dollar days and inventory days statistics).

Static buffer   That portion of the buffer which is used to protect the buffer origin but is not allowed to fluctuate.

Static data   Data which are unrelated to the changes which occur in the environment in which they are used.

Station   (*see* **Part/operation**)

Statistical fluctuation   The result of a law of nature which states that all things will vary and is evidenced in a manufacturing environment by the fluctuation in the ability of a resource to meet a specific schedule.

Subordination   Providing the constraint what it needs from other resources to maximize the amount of throughput generated.

Synchronization

T-plant   T-plants are characterized by a relatively low number of common raw material and component parts optioned into a large number of end items.

Technical implementation   That portion of the implementation dealing with hardware and software, including system selection, integration, and testing.

Temporary bottleneck   A situation created when the demand for time on a specific resource temporarily exceeds its capacity.

Tertiary constraint   The third physical constraint identified during the scheduling process.

Theory of constraints (TOC)   A management philosophy developed by Dr. Eliyahu M. Goldratt that can be viewed as three separate but interrelated areas—logistics, performance measurements, and logical thinking. Logistics includes drum-buffer-rope scheduling, buffer management, and VAT analysis. Performance measurement includes throughput, inventory, and operating expense and the five focusing steps. Thinking process tools are important in identifying the root problem (current reality tree), identifying and expanding win–win solutions (evaporating clouds and future reality tree), and developing implementation plans (prerequisite tree and transition tree) (APICS).

Throughput   One of the key measurements used to manage a TOC company. It is defined by Eli Goldratt as "the rate at which the system generates money through sales."

Throughput accounting   The use of the three basic measurements of throughput, inventory, and operating expense to manage the financial/accounting aspects of the company and make decisions.

Throughput chain   A unique chain of resources, processes, or products connected by the generation throughput (also referred to as throughput channel).

Throughput dollar days   A measurement used to minimize those activities which threaten the constraint. It is represented by the formula throughput dollar value multiplied by the number of days late.

Throughput justification The process of justifying expenditures based on their impact to throughput.

Time buffer A buffer using time as an offset for those things that will go wrong to protect the goal.

Timeline The time on the planning horizon.

Time rods A time mechanism used to separate/protect orders (batches) on schedules for different resources.

Time zero The current date and time.

Tracking zone Zone II of the buffer requiring that orders not having been received at the buffer origin before a specific time be located and tracked for possible problems.

Transfer batch size The number of parts moved from one resource to the next during production.

Transition tree (TT) Used in the TOC thinking process to define those actions necessary to achieve a goal.

Total quality management II (TQM II) The application of the focusing mechanisms and tools of the theory of constraints to the tools and methodologies provided by total quality management.

Undesirable effect (UDE) Used in the current reality tree of the TOC thinking process, the undesirable effect identifies those effects in the environment which are unwanted.

V-plant V-plants are characterized by constantly diverging operations with a small number of raw material items being converted into a large number of end items.

VAT analysis The analysis of how resources perform based on the way in which they interface with each other.

Zone I (*see* **Expedite zone**)

Zone II (*see* **Tracking zone**)

Zone III That portion of the buffer which does not include the expedite or tracking zone.

Zone profile The percentage of on-time delivery for zones I and II of the buffer.

Bibliography

Goldratt, Eliyahu M., *The Haystack Syndrome*, North River Press, Croton-on-Hudson, NY (1990).

Goldratt, Eliyahu M., *It's Not Luck*, North River Press, Great Barrington, MA (1995).

Goldratt, Eliyahu M., *Critical Chain*, North River Press, Great Barrington, MA (1997).

Goldratt, Eliyahu M., and Robert F. Fox, *The Race*, North River Press, Croton-on-Hudson, NY (1986).

Goldratt, Eliyahu M., and Jeff Cox, *The Goal*, North River Press, Great Barrington, MA (1992).

Goldratt, Eliyahu M., Eli Schragenheim, and Carol Ptak. *Necessary But Not Sufficient*, North River Press, Great Barrington, MA (2000).

Lockamy, Archie III, and James F. Cox III. *Re-Engineering Performance Measurements*, Irwin Professional Publishing, New York (1994).

Newbold, Robert C., *Project Management in the Fast Lane: Applying the Theory of Constraints*, St. Lucie Press, Boca Raton, FL (1988).

Noreen, Eric, Debra Smith, and James T. Mackey, *The Theory of Constraints and Its Implications for Management Accounting*, North River Press, Great Barrington, MA (1995).

Orlicky, Joseph, *Material Requirements Planning*, McGraw-Hill, New York (1974).

Stein, Robert E., *The Theory of Constraints: Application in Quality and Manufacturing*, Marcel Dekker, New York (1997).

Umble, Michael M., and M. L. Srikanth, *Synchronous Manufacturing: Principles of World Class Excellence*, Southwestern Publishing Co., Cincinnati, OH (1990).

Vollman, Thomas E., William L. Berry, and D. Clay Whybark, *Manufacturing Planning and Control Systems*, Dow Jones Irwin, Homewood, IL (1988).

Index